Metallurgical Processes and Production Technology

D J Davies
BSc, CEng, FIM
formerly of South Gwent
College of Further Education

L A Oelmann
MSc, CEng, MIM
Gwent College of Higher Education

Pitman

PITMAN PUBLISHING LIMITED
128 Long Acre, London WC2E 9AN

Associated Companies
Pitman Publishing New Zealand Ltd, Wellington
Pitman Publishing Pty Ltd, Melbourne

© D J Davies & L A Oelmann 1985

First published in Great Britain 1985

Text set in 10/12 pt Linotron 202 Times
printed and bound in Great Britain
at The Pitman Press, Bath

ISBN 0 273 01894 9

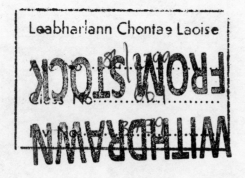

Contents

Preface

This book has been written as a companion volume to *The Structure, Properties and Heat Treatment of Metals* and completes the intended coverage of the science and technology of the common metallic materials.

In this volume the extraction and shaping of metals are considered, together with the allied topics of pyrometry, fuels and refractory materials. In addition there are sections on joining, machinability and corrosion. The subject matter has again been very concisely written with extensive use of explanatory line diagrams. Occasional reference is made to *The Structure, Properties and Heat Treatment of Metals* which deals with the more theoretical aspects of the subject.

As with the first volume this book should prove helpful to students who have embarked on B/TEC Certificate studies in Metallurgy, Cast Metals Technology, Materials Science, Mechanical and Production Engineering.

We express our gratitude to various organisations for permission to use illustrations of equipment and specific acknowledgements are made in the text. Our grateful thanks also go to Joy Davies who typed the manuscript.

DJD & LAO

1 Ore Preparation

Metals occur as chemical compounds in the earth's crust. These compounds are called *minerals*, and they are constituents of rocks. A rock which contains mineral(s) from which metal(s) may be *profitably* recovered is called an **ore**. The recovery (or "winning") of the metal is termed *extraction*, and may consist of several processing stages. The as-mined ore may require considerable preparation (or "dressing") before the extraction processes proper can begin.

1.1 Types of ore

An ore may contain three groups of minerals;

1 Valuable compound(s) of the metal being sought.
2 Compounds of other chemical elements, which may be of secondary value or which may be objectionable.
3 Minerals of no value—these are referred to collectively as the "**gangue**".

Ores may be classified according to the type of chemical compound of the valuable metal present. By far the most important groups are the **oxides** and **sulphides**. In the past, unreactive metals (e.g. Au, Ag, Pt, Cu, Hg) have been found in the elemental (uncombined) state. Such material is referred to as "native metal". However, these occurences are now rare.

Iron, aluminium, tin and titanium occur mainly as *oxide* ores or *hydrated oxides*.

The metals copper, lead, zinc and nickel occur mainly as *sulphide* ores and are usually found deep within the earth's crust and are, therefore, more costly to mine than ores nearer the surface. It is more difficult to extract the metal from sulphides than from oxides.

Oxy-salt ores include carbonates, sulphates and silicates. Some of these compounds form as a result of weathering by reaction with moisture, atmospheric oxygen or carbon dioxide, e.g. sulphates may have been produced from sulphides. Copper and zinc occur as oxy-salts as well as in the form of sulphides. These ores occur mainly near the earth's surface and sometimes even outcrop.

The *economic value* of an ore depends on many factors, including the following:

1 The extent of the deposit and its accessibility, e.g. whether the deposit is near the surface or not.
2 The form in which the metal is present and its concentration (i.e. rich or lean ore).

3 The texture of the ore, i.e. the way in which the valuable mineral is dispersed (e.g. whether aggregated in coarse particles or disseminated in fine particles).
4 The nature of the gangue.
5 The nature and concentration of impurities.
6 The nearness of the deposit to fuel, power and water supplies.
7 The demand for, and value of, the metal.

The chemical nature of the ore determines the most suitable extraction process to be applied. The impurities present, valuable or harmful, determine the details of the procedure, especially in refining, and hence influence the cost of extraction and refining. The decision to extract the metal at the mine, or to transport a concentrate for treatment elsewhere, depends mainly on the transport costs and the availability of services such as water, fuel and power.

In order to achieve *metal winning*, it is necessary to separate that component of the ore which contains the valuable metal(s) from the gangue. The first stage usually involves crushing the ore to a suitable particle size for further processing. This is followed by operations to separate as much of the gangue as possible; these operations are called *ore preparation processes*. Subsequently, other operations are performed to decompose the metal compound, separate the rest of the gangue, and obtain the metal in a relatively impure condition. The impure metal may then be *refined*.

1.2 Ore Preparation

This stage, which is also known as **ore dressing**, follows mining and prepares the ore for extraction. It controls the particle size and cleans the ore by removing much of the gangue. Dressing may also remove particular substances containing undesirable elements, which may be difficult or impossible to remove later. Hence, if ore dressing is carried out,

a) A smaller amount of material has to be taken to the extraction plant, thus lowering the transportation costs.
b) A smaller amount of material has to be used to extract a given quantity of metal, thus reducing energy costs.
c) A smaller amount of waste material is produced during extraction, thus minimising loss of metal.

The main objectives of ore dressing are

1 The **liberation** of the valuable mineral(s) from the gangue mineral(s).
2 The **separation** of the valuable mineral(s) from the gangue mineral(s).

The first objective is achieved by size-reducing processes which are referred to as *comminution*.

The second objective is accomplished by *concentration* processes which exploit differences in physical properties between the valuable and gangue minerals, such as those of relative density, magnetic and surface properties.

In addition to comminution and concentration, there are other important steps in ore dressing, such as the sizing of particles at various stages and the de-watering of pulp material. A simplified flowsheet for ore preparation is shown in fig 1.1.

Fig. 1.1 Flowsheet for ore preparation

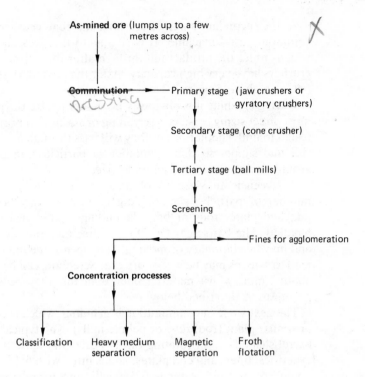

1.3 Comminution

The term **comminution** refers to various *crushing* and *grinding* processes used to decrease the particle size of the ore in order to release or expose as much of the valuable mineral as possible. Obtaining the optimum degree of liberation of the valuable mineral from the gangue minerals is of the greatest import- ance. If the particle size is made unnecessarily small, it is not only wasteful in fuel and power, but also makes subsequent separation of the gangue more difficult. During each stage of comminution, the particle size is decreased in the approximate ratio of 3 or 4 to one. At each stage, some fines are produced regardless of the reduction ratio, so that screening is used between the stages to remove the fines which would not be further reduced in subsequent comminution.

Comminution may be divided into crushing and grinding, while crushing can be sub-divided into primary and secondary stages. The primary stage is carried out in jaw crushers or gyratory crushers. The as-mined ore goes into a wedge-shaped space between a fixed crushing plate and a moving one. In **jaw crushers** (fig. 1.2) the ore is squeezed until it breaks and the fragments fall down to a narrower part of the wedge to be squeezed repeatedly, until they fall through the minimum gap at the bottom. The **gyratory crusher** (fig. 1.3) is similar in effect, but the relative motion of the crushing faces is due to the gyration of the eccentrically mounted cone. Jaw crushers can handle larger lumps (up to about 2 m across), are more flexible, and are cheaper to run and maintain. However, gyratory crushers have a higher rate of throughput. The particle size after the primary stage is between about 15 cm and 5 cm across.

In the secondary stage of crushing the **cone crusher** (fig. 1.4) is the most commonly used machine. It is similar to the gyratory crusher, but the gap widens to let the product fall through after the "bite" has been made. These crushers have very high capacity. After the secondary stage the particle size is between about 5 cm and 2 mm across.

During crushing it is obviously necessary to size the product. The simplest, most direct sizing process is **screening** or *sieving*. In principle, the particles are tested to find whether or not they will pass through an aperture of a particular size and shape, so that a number of particles can be separated into two groups—an oversize and an undersize.

A screen is an assembly of such apertures designed to test a very large number of particles simultaneously. Screens may be made of steel wire, punched plate, rods or bars depending on the size of the particles being handled. The bar screens used for coarse material are known as "grizzlies", and may be stationary or moving. Screens for fine material are vibrated. The feed to screens may be wet or dry ore. Screening can be used for sizes down to about 2 mm, whilst material finer than this may be sized by classification processes as described below.

The next stage of comminution is **grinding**, which is carried out in ball mills or similar plant (rod, tube or pebble mills). The typical ball mill (fig. 1.5) is a barrel-shaped vessel rotating on its horizontal axis. It has replaceable steel cylindrical liners and end plates, and is fitted with balls of steel or cast iron. A "pulp" of ore and water is fed axially into it and the discharged overflow carries the ground product. The speed of rotation of the mill is critical. The drum must not be rotated too quickly or else the contents will centrifuge. On the other hand, if rotation is too slow, the material will slide down the side of the drum. Neither of these conditions will result in grinding taking place. The correct speed of rotation will result in a cascading effect which will cause proper grinding to occur. Grinding reduces ore of about 5 mm particle size to the approximate range 0.5 mm to 0.05 mm.

After comminution, some of the resulting product may be too fine for efficient extraction and special agglomeration treatment may be necessary to reform the fine particles into larger lumps.

The wet processing of finely divided ore suspended in water is often followed by a **dewatering** treatment in order to lower the ratio of water to solids. The most widely used dewatering method is a continuous gravity settling process, known as *thickening*, which is a relatively cheap, high capacity process. A continuous thickener (fig. 1.6) consists of a shallow cylindrical tank, up to about 200 m in diameter and 3 m to 7 m deep, with a central drive carrying radial arms, equipped with blades, shaped so as to rake the settled solids towards the central outlet. Pulp is fed into the centre, the solids moving continuously downwards and then inwards towards the outlet, *while* the liquid moves upwards and outwards. The clarified liquid overflows while the solids are pumped away as thickened pulp from the outlet.

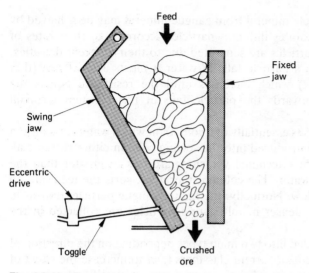

Fig. 1.2 Jaw crusher

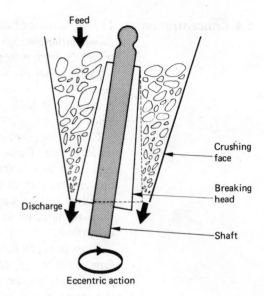

Fig. 1.3 Gyratory crusher

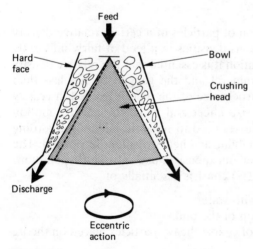

Fig. 1.4 Cone crusher

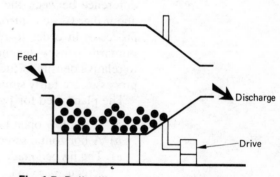

Fig. 1.5 Ball mill

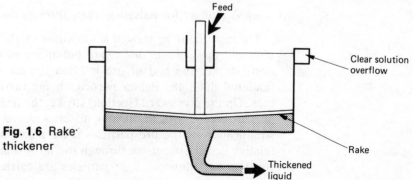

Fig. 1.6 Rake thickener

1.4 Concentration

The separation of valuable mineral from gangue mineral may be achieved by **classification**, which separates different particles according to their rates of travel through water. Particles are separated due to their different densities, sizes and shapes. When a particle falls in water it meets a resistance (that depends on its velocity) which increases until the resistance equals the gravitational pull. Afterwards the particle falls at this constant terminal velocity.

Classifying plant consists essentially of a column in which water is rising at a uniform rate. Particles introduced into the sorting column either rise or sink according to whether their terminal velocities are less or greater than the upward velocity of the water. The column, therefore, sorts the feed into an overflow and an underflow. Normally, the lighter gangue particles appear in the overflow, while the denser metal-bearing particles are collected in the underflow.

Classifiers can be divided into two main types depending on the direction of flow of the water. *Horizontal* current classifiers tend to increase the effect of *particle size* on the separation, whereas *vertical* current classifiers accentuate the effect of *density*.

Gravity Separation

This method achieves the separation of particles of a certain relative density from a mixture of particles of various densities in a feed of fairly uniformly sized particles. For effective separation it is essential that the relative density difference between the valuable mineral and the gangue be not less than about 0.5. Close control of the particle size of the feed is needed in gravity processes in order to reduce the size effect and make the particle motion primarily density-dependent. Processes used to separate minerals according to relative density include jigging, tabling and heavy medium separation. The processes are fairly simple, relatively inexpensive, and cause little pollution.

The plant used for **jigging** (fig. 1.7) consists essentially of

a) A shallow open tank filled with water.
b) A horizontal screen at the top of the tank.
c) The jig bed made of a layer of coarse, heavy particles placed on the jig screen.
d) A water compartment or hutch with a discharge spigot for concentrate removal.
e) A plunger for pulsating water through the screen.

The material to be treated is supported on the screen which is subjected to alternate upward and downward pulsations of water. On the upward (pulsion) stroke, the bed of ore is lifted against the gravitation force and is rendered fluid, the lighter particles being carried further than the heavier ores. On the downward (suction) stroke, the heavier material tends to collect on the screen, with the lighter material above it. After a few such cycles, stratification of the ore particles takes place, so that particles with a high relative density penetrate through the jig bed and screen to be drawn off as concentrate, while the light particles are carried away to be discarded, or

Fig. 1.7 Jig

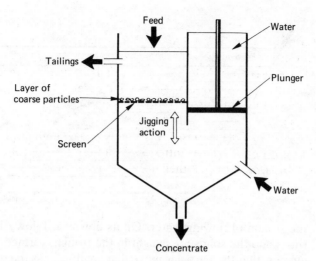

further treated. The motion produced by the eccentric drive is supplemented by a large amount of continuously supplied hutch water which enhances the upward, and diminishes the downward, velocity of the water.

The particle size range of feed to jigs varies from about 20 mm to 0.1 mm. Jigs are usually used to concentrate relatively coarse material and, if the feed is close-sized (e.g. 10 mm to 2 mm), it is easy to achieve good separation of a narrow range of relative density (about 0.5). When the difference in relative density is big, good concentration is possible with a wider range of particle sizes. Jigs do not usually produce, in one operation, both a finished **concentrate** and a discard called "**tailings**"; they often produce a concentrate plus an intermediate product, known as a "**middling**", which contains sufficient valuable mineral to merit further processing.

The principle underlying **tabling** is that, when a stream of water flows over a flat, inclined surface, the velocity of the stream is lowest near the interface with the inclined surface and increases as the surface of the water is approached. Particles of high density in the stream move more slowly than lighter ones, and small particles move more slowly than large ones. Thus, a flowing stream of water can separate large, light particles from small, dense particles. This principle is used in the shaking-table, which is an efficient type of gravity concentrator and is used to produce finished concentrates from the products of other types of gravity separators.

A shaking table (fig. 1.8) consists of a wooden deck covered with rubber, over the greater part of which there are shallow ribs or riffles. These riffles are parallel strips of wood, tapering form about 1 cm high to a thin edge at the cleaning zone. These are regularly spaced and form a number of troughs. The deck is slightly inclined to the horizontal plane (from the feed-end to the discharge-end) and is vibrated in a direction along the riffles and at right angles to the pulp flow, with a slow forward stroke and a rapid return. Pulp feed, with a particle size from about 2 mm to 0.02 mm together with wash water are introduced along the top side of the deck. The mineral particles are thus subjected to two forces at right angles due to the motion of the table and

Fig. 1.8 Table

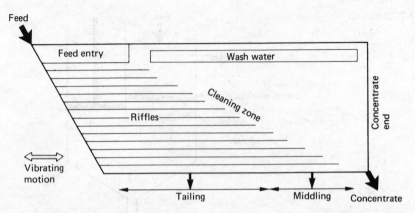

the stream of flowing water. On its downward flow, the pulp encounters the riffles and the solids are caught in the troughs formed by the riffling. The net effect is that the particles move diagonally across the deck from the feed end; the smaller, denser particles moving towards the concentrate launder at the far end, while the larger, lighter particles are carried over the riffles until they are discharged over the edge of the table into the tailings launder, which runs along the length of the table. A "middling" is also made; the middlings are affected to a greater extent than the concentrate by the flow of wash water, and are flushed down the deck and discharged from the table between the tailing and the concentrate.

Because gravity techniques may only result in a partial separation of the valuable mineral, further concentration may be required. This is achieved by using the different physical characteristics of the various particles present, such as magnetic and surface properties.

Heavy medium separation (also known as the "sink and float" method) is, in principle, the simplest of all gravity processes because it depends only on the relative density of the material. In a heavy liquid of suitable density, minerals lighter than the liquid will float, while those denser than it will sink. For example, silica has a relative density of 2.65, whereas metallic oxides and sulphides have relative densities greater than 3.5, so that silica will float and the latter will sink in a medium of relative density 3.0.

Unfortunately, there is no ready-made simple liquid of convenient density which could act as a medium. Heavy liquids usually give off toxic fumes so that the use of such substances has not proved practicable, and suspensions of finely ground solids in water are more convenient. In order to maintain continuous suspension, agitation is necessary. The solid constituent of the medium must be hard with no tendency to produce slimes. The medium must be easily recoverable from the ore particles by washing with water, and must not react chemically with the ore or corrode in any way. A suspension of ferro-silicon with a relative density of about 3.2 is commonly used.

In operation, the ore is crushed to a fairly coarse particle size, screened to eliminate fines, and fed to the separating vessel (fig. 1.9) containing the heavy medium at a pre-determined relative density. The light particles float and are removed by paddles or simply by overflow. The removal of the heavy

Fig. 1.9 Sink and float unit

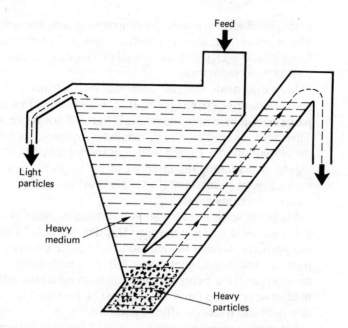

Feed

Light particles

Heavy medium

Heavy particles

particles, without causing downward currents in the vessel, is more difficult, and various methods are in use including airlift, screw-conveyor and centrifugal force.

A good degree of separation results from this method when a significant difference in relative density (not less than about 0.2) occurs at a particle size of not less than about 2 mm diameter. Separation efficiency decreases with decreasing particle size due to the slower settling rate of small particles. The method works well with ores in which the minerals are coarsely aggregated, but it is not applicable to material in which the valuable minerals are finely disseminated because, in such cases, liberation of the valuable mineral is difficult. Heavy medium separation is widely used to pre-concentrate crushed material before final grinding.

Magnetic Separation

Magnetic separation makes use of the difference in behaviour of minerals when placed in a magnetic field. Minerals can be classified according to whether they are

a) nonmagnetic (diamagnetic)
b) weakly magnetic (paramagnetic) or
c) strongly magnetic (ferromagnetic)

Magnetic separators are used to separate valuable magnetic mineral from non magnetic gangue (e.g. magnetite Fe_3O_4, from quartz), or magnetic contaminants from nonmagnetic valuble minerals.

Paramagnetic materials can be concentrated in high-intensity magnetic separators, whereas ferromagnetic materials, having very high susceptibility to magnetic forces, can be concentrated in low-intensity magnetic separators. The main ferromagnetic material separated is magnetite (Fe_3O_4) although haematite (Fe_2O_3) and siderite ($FeCO_3$) can be roasted to form Fe_3O_4 to enable good separation.

Magnetic fields may be obtained from permanent magnets, but intense fields require electromagnets. In a magnetic field of *uniform* flux, magnetic particles will orientate themselves, but will not move along the lines of flux. Therefore, as well as a high magnetic intensity, an essential requirement is the setting-up of a steep *gradient* in the magnetic field. There must also be a means of regulating the intensity of the magnetic field (in order to deal with various types of material), and this is achieved by varying the current in the case of electromagnets.

Magnetic separation is usually a continuous process and belts or drums (fig. 1.10) are used to carry the feed through the field. This enables the speed of the particles through the magnetic field to be closely controlled. The gangue particles are nonmagnetic and fall off immediately, whereas the valuable mineral particles, being magnetic, remain under the influence of the magnetic field to be released later. Means must be provided to collect the magnetic and nonmagnetic fractions after separation.

Magnetic separators can be divided into high-intensity and low-intensity machines which can be further sub-divided according to whether the feed is wet or dry.

Low intensity machines are suitable only for strongly magnetic materials. Dry separation is carried out on material with a particle size larger than about 50 mm. Wet separation is used on finer particles, resulting in less dust loss and a cleaner product because water causes a better dispersion and presents the feed more efficiently.

High intensity separators are used to treat weakly magnetic ores. Dry separation can be used on coarser material, but air currents and adhesion of particles have adverse effects on fine ores. The biggest recent improvement in magnetic separation has been the development of high intensity wet machines. Increasing use is being made of this method for treating finely ground haematite ores.

Froth Flotation

In this method, finely ground ore (particle size 0.2 mm to 0.01 mm) is placed in a tank of liquid through which air is bubbled. Surface tension effects are used to float off the particles of the valuable mineral to the surface of the liquid, while the gangue particles fall to the bottom. For flotation to take place, an air bubble must attach itself to a particle of valuable mineral and lift it to the surface of the water. Therefore the mineral particles must be very small, and must also be made water-repellent (hydrophobic). On reaching the surface, the air bubbles must continue to support the mineral particles, i.e. a stable froth must be formed. To obtain these conditions, certain flotation reagents must be present in the liquid in the tank; these reagents include frothers, collectors and various modifying agents.

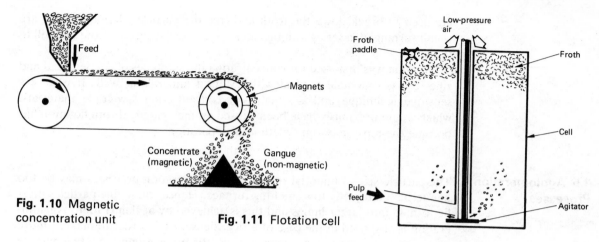

Fig. 1.10 Magnetic concentration unit

Fig. 1.11 Flotation cell

Frothers are added to stabilise the air bubbles in the liquid, so that the bubbles do not burst even on reaching the surface of the liquid. A froth should be produced which is just stable enough to carry the load of floatable mineral into the collecting launder. Excessive frothing must be avoided. Frothers are usually organic reagents (e.g. pine oil) capable of being adsorbed on the air-water interface.

Collectors are organic compounds which make selected minerals water-repellent (hydrophobic) by forming a very thin non-wetting layer on the particles. Such particles will then be floated to the surface by attachment to small air bubbles introduced into the water. Typical collectors include xanthates, thiocarbonates, dithiophosphates for floating sulphides; fatty acids and soaps for separating oxides and amines for collecting silicates. **Modifying agents** are used to increase the selectivity of the process as follows.

Activators are added to render certain minerals hydrophobic (and thus floatable) which otherwise would not float.

Depressants are used to make certain minerals hydrophilic (wetted by water), thus preventing their flotation.

Regulators are used to control the pH value of the solution, preventing the precipitation of compounds which might adversely affect the separation. Flotation is usually carried out in alkaline solutions, and the alkalinity is controlled by additions of lime, sodium carbonate or sodium hydroxide if the pH needs increasing and by sulphuric acid if the pH needs lowering.

Control of pH with judicious choice of activator or depressor enables separation to be made of two or more valuable minerals from one another, and from the gangue, hence recovering separately the valuable constituents from a complex ore. This is known as selective or differential flotation.

The reagents are often blended into the pulp by agitation in "conditioning" tanks before being fed into the flotation cells. In the cell (fig. 1.11) the pulp is agitated and air is injected to form a large number of small bubbles which rise through the cell, collecting mineral particles as they rise. They collect as a froth at the surface and this is scraped off into a launder where water sprays are used to break down the froth and free the particles. Flotation cells are usually arranged in series with the liquid continuously passing from one cell to the next.

Flotation was first used for concentrating the sulphides of copper, lead and zinc, but is now able to deal with oxidised minerals as well. It is a very sensitive technique and is capable of concentrating low-grade materials, which would previously have been classed as uneconomic. Froth flotation has become the most important method of separation.

1.5 Agglomeration Processes

The particle size of material resulting from separation processes may be too fine to be fed directly to a smelting furnace. Hence, these fine particles must be formed into larger lumps and this is achieved by **agglomeration**. Agglomeration is also used in the case of a rich, dense ore in order to make it more amenable to reaction with a reducing gas; the ore is ground to a fine particle size and the particles then reformed into larger, porous lumps more suitable for smelting.

The advantages of agglomeration are

1 Fine material (including flue dust) is reclaimed.
2 Improved reducibility results in increased production rate.
3 More uniform working of the smelting furnace is achieved.
4 The addition of a flux (e.g. limestone in iron smelting) to the mixture to be agglomerated will give a self-fluxing product.

The main methods of agglomerating ore fines are sintering and pelletising.

Sintering is carried out at elevated temperatures when incipient melting of the charge occurs. Particles of ore about 5 mm diameter are mixed with about 5% coal dust or coke breeze (to generate heat) and about 10% water (to improve the permeability of the sintered product). Sintering is a continuous process and is usually carried out in a Dwight-Lloyd type of machine (fig. 1.12). The damp mixture of ore fines and coke breeze is fed into cast iron pallets on an endless belt. The charge is passed under an ignition hood (fired by gas or oil) which ignites the coke in the mixture. Suction fans draw air down through the charge as it passes over several wind boxes. After being cooled by the last wind box, the sinter is broken up into conveniently sized lumps.

Sintering of the charge takes place in two stages:

1 Drying and some chemical decomposition occurs with the removal of H_2O, CO_2 and SO_2. This stage is called calcination, during which any voltaile elements present (e.g. Zn, Cd, As) may also be driven off and recovered.
2 Incipient melting takes place and the liquid formed flows by capillary action into spaces between the ore fines, thus strengthening the sintered product.

Sintering is also used for dealing with sulphides, but in this case the main objective is to convert the sulphide into oxide, so that it is most important to avoid incipient fusion of the sulphide particles, otherwise oxidation would

Fig. 1.12 Dwight-Lloyd sintering machine

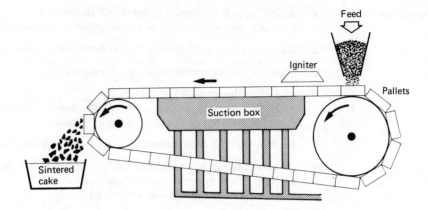

cease. The plant used is similar, except that air is blown upwards through the bed in order to allow easier collection of SO_2 as a byproduct.

A disadvantage of sintering is that the process is thermally inefficient; large volumes of air must be heated and drawn through the sinter bed and, because the issuing gases are laden with dust, their heat cannot be readily recovered. The discharged sinter may be at a temperature of over 1000°C, but it cannot normally be charged to the smelting furnace until it is cooled.

If the particle size of the ore fines is less than about 1.5 mm, sintering becomes unsuitable because of the risk of complete fusion of such small particles. Such particles are agglomerated by **pelletising**, which is carried out at room temperature with the use of a binding agent.

The ore fines are mixed with about 10% water (to improve permeability) and a binding agent, which may be organic material, lime or a silicate depending on the type of ore. The mixture is fed into an inclined rotating drum and the particles bond together to form pellets about 10 mm to 25 mm in diameter depending on the time in the drum. The optimum size of pellet depends on the ore composition and subsequent treatment. The pellets discharged from the drum are too soft to withstand the loads encountered in a tall furnace (such as a blast furnace), so that they need to be hardened by heating at about 1000°C; this firing may be carried out in a Dwight-Lloyd type of plant or alternatively in a small shaft-type furnace.

Pelletising results in a product which is more easily reduced than sinter because the surface of the pellets show greater porosity. Also the very small particles arising from modern separation processes are better dealt with by pelletising than by sintering.

Exercises 1

1. a) Distinguish between a mineral and an ore.
   b) Name two types of chemical compound of the valuable metal in an ore.

2. Name four factors governing the economic value of an ore deposit.

3. State the two main objectives of ore preparation.

4. a) Define the term "comminution".
   b) Name one type of plant used for each stage of comminution.
   c) Make a sketch of one type of comminution plant.

5. Explain why the comminution process is necessary in ore preparation.

6. a) Define the term "concentration" as applied in ore dressing.
   b) Name three types of concentration process.
   c) Describe one type of concentration process.

7. Write a brief account of each of the following:
   a) jigging
   b) tabling.

8. a) Explain what is meant by the term "heavy medium separation".
   b) Make a simple sketch of one type of plant used for heavy medium separation.

9. Write an account of the froth flotation method of ore concentration.

10. Explain what is meant by "differential flotation".

11. a) Define the term "agglomeration".
    b) Name two advantages of agglomeration.
    c) Name two methods of agglomeration.
    d) Make a sketch of one plant used for agglomeration.

2 Extraction and Refining of Metals

Extraction processes may be divided into three main groups:

1 *Pyrometallurgical* methods in which heat is used to achieve decomposition of the valuable metal compound.
2 *Hydrometallurgical* methods which involve taking the valuable metal into aqueous solution followed by its recovery from the solution.
3 *Electrometallurgical* methods in which electrolysis is used to decompose a fused compound of the valuable metal.

The choice of extraction process is largely governed by the type of ore and the chemical stability of the valuable minerals present. However, the details of the procedure will also depend on the energy costs, production rate needed, and the required metal purity. Hitherto, the pyrometallurgical route has been predominant because of the abundance and lower cost of fossil fuels and also due to its suitability for high production rates. The emphasis on the control of atmospheric pollution has, however, resulted in a trend towards more hydrometallurgical processing, especially if hydro-electrical power is available. Hydrometallurgical extraction is slower than pyrometallurgical methods, but it is more suitable for dealing with lean ores.

2.1 Pyrometallurgical extraction

Pyrometallurgical extraction processes involve the application of high temperatures to achieve chemical reduction of a metal compound to the metal. The metal compound used is usually an oxide and the high temperatures are generally produced by combustion of coke, oil or gas in a furnace. Electricity is also sometimes used. The reduced molten metal and a molten slag (composed of the unwanted gangue oxides) are produced as separate layers. In addition to ore reduction, pyrometallurgy may include preliminary operations such as drying, calcination, roasting and sintering. These processes result in chemical and/or physical changes which make the ore more amenable to furnace smelting.

The mechanical process of **de-watering** by thickening and filtration removes only part of the water from a material, whereas **drying** refers to the removal of all non-combined water. A current of hot air or gas is passed through and over the material, causing water to be removed by evaporation. **Calcination** involves heating to rather higher temperatures and removes combined water and volatile constituents. Carbonates are decomposed with the evolution of CO_2 (e.g. $CaCO_3 \rightarrow CaO + CO_2$) while hydroxides may be changed into oxides (e.g. $Ca(OH)_2 \rightarrow CaO + H_2O$). Drying and calcination are carried out in rotary kilns, shaft furnaces or fluidised bed furnaces.

Roasting consists of heating ore below its fusion point with access of air in order to cause chemical change (usually oxidation) to take place so that the product is more amenable to subsequent processing. The most important roasting reactions are those involving sulphides, typified by the following:

$$2MS + 3O_2 \rightarrow 2MO + 2SO_2 \quad \text{(dead roast)}$$
$$MS + 2O_2 \rightarrow MSO_4 \quad \text{(sulphating roast)}$$
$$\text{or} \quad 2MO + 2SO_2 + O_2 \rightarrow 2MSO_4 \quad \text{(sulphating roast)}$$

Dead roasting is carried out at about 800–900°C with excess air and is used for producing an oxide which is to be subsequently reduced by carbon or hydrogen. A *sulphating roast* is achieved at about 600–700°C, with a restricted amount of air and is used to produce a material which can be leached with dilute H_2SO_4 to give a solution from which the metal may be readily recovered. A less-common type of roasting reaction may result at high temperatures where oxide and sulphide interact to produce the metal—known as a *reduction roast*:

$$2MO + MS \rightarrow 3M + SO_2$$

Another type of roast is the so-called *"magnetising" roast* in which Fe_2O_3 is reduced to Fe_3O_4, which can then be magnetically separated.

During the early stages of a roasting operation drying and calcination will take place as previously described. Volatile impurity oxides (e.g. As_2O_3, Sb_2O_3) may be expelled. Many roasting reactions are exothermic, so that once the reaction has been started the liberation of heat enables the reaction to continue spontaneously. This is called *autogenous roasting* and requires careful control, otherwise overheating will cause fusion and consequent stopping of oxidation. Autogenous roasting does not usually result in a completely dead roast and, to produce material with very low sulphur content, external heating is necessary.

Roasting reactions are examples of gas-solid reactions and rely on the diffusion of oxygen into, and SO_2 out of, each particle of ore. The roasted product, known as calcine, must be chemically and physically suited to the next stage of the extraction process. Thus, the calcine should be in a granular form if it is required as feed material to the blast furnace, whereas fine material may be needed in other cases. In order to meet the different requirements, various types of roasting plant are in use.

To produce granular calcine sintering (see section 1.5) is suitable, and the rapid air movement involved obviously increases reaction rates. However, in the treatment of sulphides (where a dead roast is a primary aim), it is essential that premature incipient melting be avoided, otherwise oxidation will cease. To achieve this condition it may be necessary to carry out the sintering in two stages—the charge for the second stage consisting of some already sintered material, together with raw sulphide concentrate.

In **multi-hearth roasting** furnaces (fig. 2.1), the ore fines are passed from the top of the furnace through a series of flat hearths to ground floor level, and a counter current of hot air passes upwards. The product consists of finely divided calcine suitable for hydrometallurgical processing (see section 2.2).

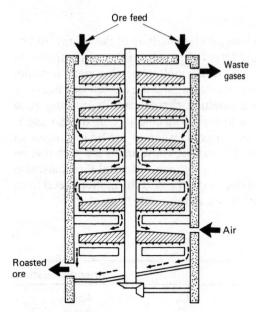

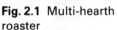

Fig. 2.1 Multi-hearth
roaster

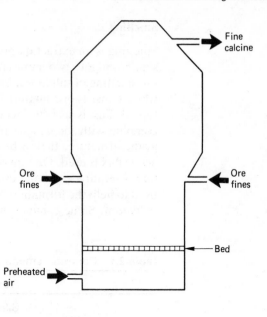

Fig. 2.2 Fluo-solids
roaster

Because of the build-up of accretions on the hearths and rabbles of these furnaces, the maintenance costs are high, so that multi-hearth roasting is being superseded by fluo-solids roasting.

In the **fluo-solids roaster** (fig. 2.2), pre-heated fine ore (particle size less than about 5 mm) is injected through a burner into a reacting chamber and fluidised by a current of air pre-heated to about 500°C. At a critical air velocity, the frictional forces exerted on the ore particles begins to exceed their weight, lifting occurs, and the particles become mobile within the ore bed which begins to resemble a liquid of high viscosity. With increase in air velocity, movement of the particles becomes rapid and the bed assumes a highly turbulent state. Continuous circulation of ore particles occurs and an almost uniform composition is reached. Fluidisation of the bed depends mainly on particle size and air velocity, but particle shape and density also play a part.

Fluo-solid furnaces contain no internal moving parts so that maintenance costs are low. The resulting product is suitable for hydrometallurgical extraction, but the finely divided state of the calcine precludes its use for blast furnace smelting unless some form of briquetting is used.

The sulphide-sulphur content is decreased to about 0.1–0.2%, with a sulphate-sulphur content of 1–2%.

Smelting

Smelting is an extraction process which involves melting of the charge and its separation into two immiscible liquid layers of metal and slag respectively. In the smelting of sulphides, instead of producing molten metal, a **matte** is made, which consists of a mixture of molten sulphides, and the molten slag floats on top. A flux is usually included in a smelting charge, and its purpose is to combine with the gangue of the ore to form a fusible slag. If the gangue is acidic in nature, then a basic flux is required, whereas for basic gangue an acidic flux is used. On rare occasions, the gangue constituents are such that no flux is required, and the ore is said to be *self-fluxing*. Neutral fluxes are also used to help the formation of a fluid slag and to protect the molten metal from oxidation. Some common fluxes are listed in table 2.1.

Table 2.1 Common smelting fluxes

Flux		Chemical nature
Silica	SiO_2	Acid
Lime	CaO	Basic
Iron oxide	FeO	Basic
Fluorspar	CaF_2	Neutral

Metal oxides are smelted under highly reducing conditions usually with carbon; reduction with hydrogen is less common, whilst reduction with another metal which forms a more stable oxide is occasionally carried out.

In general terms, metal oxide smelting can be represented as:

$$\text{Metal oxide} + \text{Reducing agent} + \text{Flux} + \text{Heat} \rightarrow \text{Metal} + \text{Slag} + \text{Gas}$$

It is seen, therefore, that metal oxide smelting (unlike matte smelting) is a main extraction process. However, the metal produced is impure and may require a further refining treatment.

Matte smelting is carried out under neutral or slightly reducing conditions and, because metal sulphides have lower melting points than oxides, the process is conducted at lower temperatures than oxide smelting. A considerable amount of the impurities separates into the slag, leaving the matte richer in the metal being sought. Therefore matte smelting is a concentrating stage in the extraction of a metal from its sulphide ore.

Matte smelting used to be carried out in reverberatory furnaces (fig. 2.3) which have a rather shallow hearth with an arched roof, from which the hot combustion gases are deflected (reverberated) on to the charge. The long, wide hearth also allows good separation of the molten matte and molten slag. However, high consumption of hydrocarbon fuel and air-pollution problems due to the sulphur dioxide evolved is causing the reverberatory furnace to be superseded by flash-furnace smelting, which also achieves a higher rate of production.

Fig. 2.3 Reverberatory furnace

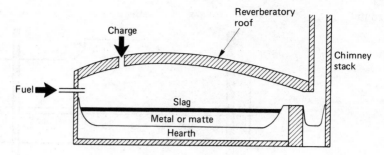

Fig. 2.4 Flash-furnace smelter

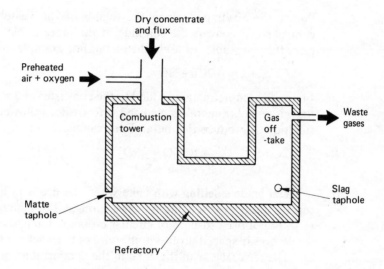

In **flash-furnace smelting** (fig. 2.4), dried, finely ground sulphide concentrate, together with flux, is injected into a hearth-type furnace where it is combusted with pre-heated oxygen-enriched air. The fine particle size allows large contact area between the concentrate, flux and oxidising atmosphere, so that a high throughput is possible. The process will be autogenous if the amount of sulphur present generates sufficient heat (from the exothermic oxidation reactions) to obviate the need for using fuel.

When higher processing temperatures (in excess of 1500°C) are needed, electric arc furnaces may be used. Also, on the infrequent occasions when the concentrate is in lump form, a blast furnace may be used for matte smelting, but highly reducing conditions are not required so that the amount of coke charged is limited.

Matte conversion to the metal is achieved by blowing air, or oxygen-enriched air, through the molten matte, thus causing selective oxidation of the more reactive impurity sulphides to their respective oxides which are transferred to the slag. The process is carried out in a horizontal converter (fig. 2.5) lined with a basic refractory. Molten matte is poured into the converter and the air for oxidation is introduced through openings, known as

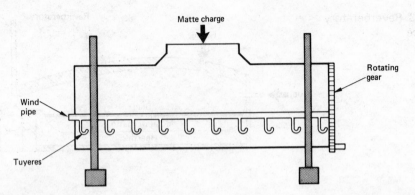

Fig. 2.5 Horizontal converter

"tuyeres", positioned along the length of the vessel. The air-blowing is controlled to convert the sulphide of the more noble metal present (which gives the less stable oxide) to the metal. For example, in the case of copper,

$$Cu_2S + O_2 \rightarrow 2Cu + SO_2$$

In practice, more matte is added to the converter and some of the sulphide of the more noble metal is changed into oxide, followed by interaction with sulphide to produce the metal. For example,

$$2Cu_2S + 3O_2 \rightarrow 2Cu_2O + 2SO_2$$
$$2Cu_2O + Cu_2S \rightarrow 6Cu + SO_2$$

Metal oxide smelting with coke when the ore is in lump or agglomerated form is usually carried out in a blast furnace. For higher melting point metals (e.g. Fe), a blast furnace of circular cross-section is used (fig. 2.6) which has small, evenly spaced tuyeres at the base of the stack for the entry of preheated air. The air reacts with the coke in the combustion zone to produce CO:

$$2C + O_2 \rightarrow 2CO$$

Some CO_2 is formed, but this reacts with the excess carbon to form CO:

$$C + O_2 \rightarrow CO_2 \qquad CO_2 + C \rightarrow 2CO$$

Reduction of metallic oxide begins in the upper part of the stack where CO gas acts as the reducing agent. This is referred to as *indirect reduction*. At the higher temperatures lower down the furnace stack, *direct reduction* by the coke takes place. The heat required for reduction to take place is generated by combustion of the coke, which also has to give support to the furnace charge. The ore, coke and flux are charged into the top of the furnace. The hot gases, rich in CO, rise up the stack, heating and reacting with the charge descending towards the hearth. These opposing movements constitute a counter-current process, which affords excellent contact between the reactants, thus resulting in good rates of reaction. The nitrogen in the ascending gases, although it plays no part chemically, acts as a heat transfer medium helping to heat the descending charge.

In the case of metals with relatively low boiling points (e.g. Zn, Pb), very high furnace temperatures must be avoided because of the danger of metal

Fig. 2.6 Blast furnace
of circular section

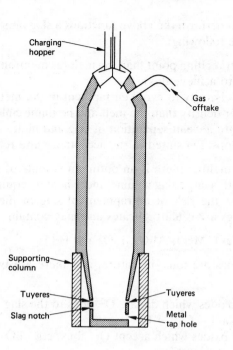

Fig. 2.7 Blast furnace
of rectangular section

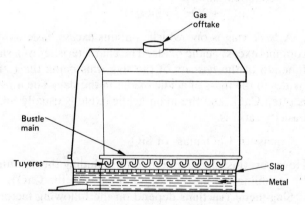

volatilisation. Therefore a blast furnace of rectangular cross-section (fig. 2.7) is used and the tuyeres are placed only along the long sides. This arrangement results in lower furnace temperatures.

Slags play an important part in all metal smelting processes and the slag composition must ensure satisfactory composition of the metal (or matte) being produced. The *functions* of a slag in molten metal processing are as follows:

1 To protect the melt from contamination from the furnace atmosphere.
2 To accept unwanted impurities present in the melt.
3 To control the supply of refining agents to the melt through additions to the slag.
4 To insulate the melt.

In order to perform the above functions a slag must have certain *properties* including the following:

a) A lower melting point than the melt (as mentioned above, fluxes may be added to achieve this).
b) A low viscosity to avoid entrapment in the metal.
c) A lower density than the melt and be immiscible with the melt in order to ensure a clean separation of slag and melt.
d) A composition suited to the acceptance and retention of impurities.

In metal smelting there is an optimum volume of slag to achieve efficient working. With a large slag volume, more heat is required and energy costs are higher, while the risk of entrapment of slag in the metal is also greater. Smelting slags are usually silicates and may contain

$$SiO_2, \; CaO, \; MgO, \; Al_2O_3, \; P_2O_5 \; and \; FeO$$

Molten slags are ionic in nature, and there are two main types of oxide present:

1 Basic oxides which donate O^{2-} ions to the slag, e.g. CaO, MgO, FeO.
   $$(CaO \rightarrow Ca^{2+} + O^{2-})$$
2 Acidic oxides which accept O^{2-} ions, e.g. SiO_2, P_2O_5, Al_2O_3 and form complex anions.
   $$(SiO_2 + 2O^{2-} \rightarrow SiO_4^{4-})$$

A *basic* slag is one which contains excess basic oxide, while an *acidic* slag contains excess acidic oxide. The characteristics of a slag are to some extent a function of the *basicity* of the slag, this being the ratio of the mass of basic oxides to the mass of acidic oxides in the slag. The most important basic oxide is often CaO, and the main acidic oxide is usually SiO_2, so that the simplest basicity ratio is

mass of CaO/mass of SiO_2

However, this can be modified by including the amounts of other oxides (e.g. by including the mass of MgO along with the CaO).

Slag-metal reactions depend on the following factors:

1 The composition of the metal and the slag.
2 Temperature.
3 Time of contact.
4 Area of contact between metal and slag.
5 Degree of turbulence (mixing) in the two layers.

Control of the slag composition is of the greatest importance in influencing slag-metal reactions. The slag must be sufficiently fluid to present a suitable interface and to flow freely from the furnace. Fluidity increases with rising temperature, but in order to restrict fuel costs, a compromise has to be reached and slag temperatures limited.

Impurities in a metal can be removed into the slag by choosing a slag composition which has chemical affinity for the impurities. For example, an oxide impurity which has *acidic* properties will react with a *basic* slag, but not with an acidic slag. *Amphoteric* oxides (e.g. ZnO, PbO, SnO$_2$) behave as acidic oxides towards basic slags and as basic oxides towards acidic slags, so that these oxides may be removed by either type of slag. In some instances Al$_2$O$_3$ may also behave as an amphoteric oxide.

Slags may also have oxidising or reducing properties. An *oxidising slag* is one which is able to transfer oxygen atoms (not ions) to the melt in order to refine the metal by oxidising impurities. A *reducing slag* is not able to transfer oxygen atoms to the melt, but carbon is added to the slag to achieve reducing conditions. For example, powdered coal is added to a basic slag to help the removal of sulphur from iron during electric arc steelmaking.

Pyrometallurgical Refining (Fire Refining)

The impurities present in primary extracted metal may be other metals, non-metals or metalloids; they may be dissolved in, or combined with the, basis metal or sometimes mechanically entrapped. The impurites may come from the ore, fuel, flux, refractories or the furnace atmosphere. Since impurities usually alter the properties of the metal, sometimes detrimentally, it is necessary to remove or control them within certain limits.

One method of removing impurities is by preferential oxidation of more-reactive elements from a less-reactive basis metal. Such oxidation of reactive impurities may be achieved by blowing air or oxygen into the melt or, less drastically, by adding to the slag an oxide of the metal being refined. In either case, the oxide of the metal being refined, being less stable, dissociates and releases oxygen atoms to the melt. These oxygen atoms combine with impurity elements to form oxides which are insoluble in the melt. Oxides (and other insolubles) separating from a molten metal are sometimes referred to as *dross*.

Fire refining results in a metal purity of up to 99.5% and is useful only for Fe, Cu, Pb, Sn and more noble metals.

2.2 Hydrometallurgical Extraction

In general terms, **hydrometallurgical extraction** may consist of the following steps:

a) Comminution of the starting material (e.g. ore, concentrate or process waste) to a suitable particle size.
b) Roasting to convert the material into a soluble form, e.g. a sulphating roast to convert a sulphide into sulphate.
c) Leaching to dissolve the valuable metal in a leach liquor.
d) Separation of the leach liquor from the unwanted material.
e) Recovering the valuable metal from the leach liquor.

Hydrometallurgical methods may be applied to

1 Low-grade ores which are not amenable to concentration or suitable for smelting.
2 High-grade concentrates which are roasted, then processed in aqueous solution, thereby providing an alternative path to pyrometallurgical extraction.

Much of the world output of metals is produced by hydrometallurgical extraction. For example, at the present time (1982), about 75% of the world output of zinc and about 15% of the world output of copper are produced hydrometallurgically. Because of the need for stricter compliance with more stringent air-pollution legislation, the present trend is towards far more hydrometallurgical processing, and away from pyrometallurgical methods with their fume and grit problems. However, water-pollution control, aimed at reducing suspended and dissolved solids in plant effluents, also places a responsibility on hydrometallurgical plants.

A general description of the stages involved in hydrometallurgical processing is given below.

Leaching

The leaching of material of fine particle size involves dissolving the valuable metal in a suitable reagent (leachant). A leachant should be readily available in quantity, relatively cheap, must not react unduly with gangue minerals, and must be recoverable from the solution. Some typical leachants are listed in table 2.2.

Table 2.2 Common leaching reagents

Mineral	Leachant
Oxides	dilute H_2SO_4
Sulphates	dilute H_2SO_4 or water
Sulphides	$Fe_2(SO_4)_3$ solution
Alumina	NaOH solution
Cu/Ni compounds	$(NH_4)_2CO_3$ solution

Several leaching methods are used depending on ore composition, required leaching rate, and subsequent recovery and extraction methods:

1 *In situ leaching* of low-grade surface deposits, or worked-out underground mines, without removal of the ore from the mine.

2 *Heap leaching* of lumps (about 20 cm diameter) of low-grade surface deposits or mine wastes is carried out on extremely large dumps of material (e.g. 10^6 tonnes) in the open.

In methods **1** and **2**, the leachant is sprayed over the ore and trickles through the bulk of material along culverts, until it reaches a prepared impervious base where it is collected. Very long times (years) are involved in obtaining substantial solution of the valuable metal and the recovery rate is low.

3 *Percolation leaching* is carried out on ore of particle size about 5 mm in large vats. The ore is percolated by leachants of increasing concentration. A solution of sufficient concentration for economic recovery is obtained in a shorter time (days).

4 *Agitation leaching* uses finer ore (particle size about 0.5 mm or less) which is suspended in the leachant and stirred by compressed air. The process may be speeded up by making the ore particles move one way and the leachant flow in the opposite direction, thus resulting in a counter-current process and decreasing the leaching time to hours.

5 *Pressure leaching* is carried out at elevated temperature in reaction vessels called autoclaves. The solution rate is higher and the process is used to obtain solutions of minerals which are difficult to dissolve at atmospheric pressure. An example of this type of leaching is the Bayer process for the production of alumina, in which impure bauxite ore is treated with strong NaOH solution (30%) at 150–170°C under 5–10 atmospheres pressure.

6 *Bacterial leaching* is used to increase the rate of solution of certain sulphides (e.g. copper sulphide). In this case, bacteria are used to increase digestion (dissolving) rate of the valuable mineral.

The **separation of leach liquor** may be achieved by decantation, thickening or filtration. If the unwanted, insoluble solids settle readily, then the clear solution may be drawn off from the top of the tank by a pipe, which is supported by a float just below the surface of the liquid. The descent of the pipe is stopped when the sediment is reached, more slurry is added, and the process is repeated. This method, known as *decantation*, is a batch process and is slow in operation.

Fig. 2.8 Rotary-drum vacuum filter

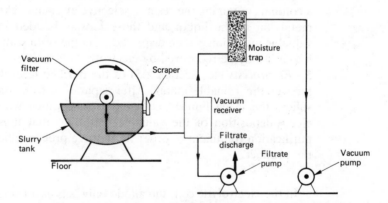

A second method of separation is called *thickening*, which is a mechanised, continuous process of combined settlement and decantation (see section 1.3). *Filtration* is used to separate fine waste solid material from the valuable leach liquor. The rotary-drum continuous vacuum filter (fig. 2.8) is the most widely used type in hydrometallurgical processing. This type of filter consists of a cylindrical drum mounted horizontally and tightly covered with a filter cloth made of natural or synthetic fibre which will not allow passage of solids. The lower half of the drum is submerged in an open tank into which the slurry to be filtered is fed. The drum rotates slowly in the feed tank and, on the application of vacuum, the pulp adheres to the filter cloth while the liquid is sucked through and drawn into the receiver, from which it is pumped away. Solution is prevented from entering the vacuum pump by the moisture trap. The filter cake is removed by a scraper.

Recovery of the Valuable Metal

Methods of recovery of the valuable metal from the leach liquor include the following:

1 *Displacement* by another metal, e.g. the addition of scrap steel to a solution containing copper ions causes precipitation of metallic copper. (This is known as the "cementation" of copper from solution.)

2 *Extraction by an organic solvent.* Certain organic liquids (e.g. ethers, esters and ketones) readily dissolve certain complex metal ions. Conditions are adjusted to make the solvent specific towards one metal ion which is removed from the aqueous solution into the immiscible layer of organic solvent. Consequently the valuable metal is concentrated into a much smaller volume of liquid from which it may be more easily recovered.

This technique is being increasingly used in hydrometallurgy for purifying solutions and for concentrating leached metals into smaller volumes of solution. (The technique is used in the extraction of uranium and zirconium.)

3 *Precipitation* of the required metal may be achieved by adjusting the pH of the solution or by passing in a reducing gas (e.g. hydrogen) under controlled conditions.

4 *Ion exchange* resins may be used to remove the required metal from aqueous solution. These resins are organic materials on which either anions or cations have been loosely bonded. When the leach liquor is passed through a column containing the resin, a selective exchange takes place of one type of metal ion in the liquor and those loosely bonded ions in the resin. The valuable metal ion is then displaced from the resin with a cheap solvent, from which it may be recovered by electrolysis.

5 *Electrolysis* (see section 2.3) of the purified leach liquor may be used to recover the valuable metal. After separation from the unwanted insoluble solids, the leach liquor may contain ions which would interfere with the electrodeposition of the valuable metal, so that it is given a preliminary purification treatment in order to improve process efficiency or the purity of the product.

In the electrolytic cell, the anode only acts as a lead-in for the d.c. current and plays no chemical part in the electrolysis; such an anode is known as an inert (or insoluble) anode and is often made of lead or carbon. The valuable metal is plated out of solution on to the cathode, which is usually made of the same metal.

No attempt is made to decrease the valuable metal ion concentration to a very low value because impurities would then be deposited as well. When the concentration becomes critically low, part of the solution is returned to the leaching plant and replaced with new leach liquor. In general, preferential deposition of the valuable metal can be obtained provided that the relevant salt of the valuable metal has a lower decomposition voltage than that of impurity salts.

Recovery of a metal by electrolysis is known as *electro-winning*.

2.3 Electrometal-lurgical Extraction

Electrometallurgy involves the electrolysis or decomposition of either an aqueous solution or a fused salt by the passage of an electric current. The two metal (or carbon) plates by which the current enters and leaves the liquid (electrolyte) are called electrodes. When no current is passing, the ions are moving at random in the electrolyte. During the passage of the current, electrons flow from the d.c. source to one electrode, called the *cathode*, which thus becomes negatively charged and attracts to itself positively charged ions (cations), e.g. metal ions. These positively charged ions consume electrons at the cathode and are made electrically neutral, thus becoming ordinary atoms. The negatively charged ions (anions) migrate to the other electrode, called the *anode*, where they lose their electrons to become ordinary atoms and the electrons then flow back to the d.c. source. The current is therefore carried through the electrolyte by a flow of ions. Figure 2.9 shows the electrolytic cell.

Fig. 2.9 Electrolytic cell

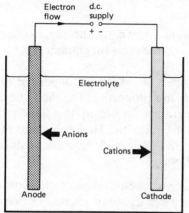

A cathode, as a source of electrons, is equivalent to a reducing agent, and discharge of positively charged ions at a cathode is, chemically, a reduction process. The hydrogen ion, for example, is reduced (by electron gain) to the hydrogen atom: $H^+ + e \rightarrow H$

Similarly, discharge of metallic ions at a cathode involves reduction to the ordinary atomic state of the metal, e.g.

$$Cu^{2+} + 2e \rightarrow Cu$$

Correspondingly, an anode, as an electron acceptor, is equivalent to an oxidising agent. Discharge of negatively charged ions at an anode is an oxidation process. For example, a chloride ion Cl^- is oxidised to the chlorine atom by electron loss, the electron being accepted by the anode and "pumped away" by the current:

$$Cl^- - e \rightarrow Cl$$

The reactions occuring at both the anode and the cathode must involve the same number of electrons. The reactions are hence chemically equivalent and involve masses in the ratio of the chemical equivalent:

$$\text{Chemical equivalent} = \frac{\text{Relative atomic mass}}{\text{Valency}}$$

Faraday's Laws of Electrolysis

Faraday carried out quantitative experiments on electrolysis and summarised his results in two laws:

1 The mass of a substance discharged (deposited or liberated) at an electrode is directly proportional to the quantity of electricity passed, measured in coulombs (1 amp passed for 1 sec = 1 coulomb):

$$m \propto It$$

where m = mass discharged (grammes)
I = steady current (amps)
t = time (secs).

or $m = kIt$

where k is a constant known as the electrochemical equivalent of the substance, and is clearly the mass (in grammes) set free by the passage of one coulomb of electricity.

2 The masses of different substances deposited or dissolved by the same quantity of electricity are proportional to their chemical equivalent weights. The quantity of electricity known as the faraday (96 487 coulombs) will discharge one gramme equivalent. Hence the mass of substance (in grammes) discharged by a current of I amps flowing for t secs (assuming no losses) may be calculated as follows:

1 faraday discharges 1 g equivalent of element
96 487 coulombs discharge A/n grammes of element

$$\therefore \quad It \text{ coulombs discharge } \frac{AIt}{96\ 487n} \text{ grammes of element}$$

where A = relative atomic mass of element
n = valency I = current

The second law may be expressed in an alternative way by stating that the quantity of electricity required to liberate 1 mole of atoms of any element is a whole number of faradays (1, 2, 3, etc.), remembering that 1 faraday represents 1 mole of electrons.

For example

1 mole of Ag atoms is liberated by 1 faraday
    ($Ag^+ + e \rightarrow Ag$)
1 mole of Zn atoms is liberated by 2 faradays
    ($Zn^{2+} + 2e \rightarrow Zn$)
1 mole of Al atoms is liberated by 3 faradays
    ($Al^{3+} + 3e \rightarrow Al$)

Using Faraday's laws, the theoretical mass of substance discharged by a known current in a given time can therefore be calculated. In practice, the actual amount of substance discharged is less than the theoretical and the current efficiency is defined as follows:

$$\text{Current efficiency} = \frac{\text{Actual mass discharged}}{\text{Theoretical mass}} \times 100\%$$

Electrometallurgical processes can be divided into:

1 Those conducted at high temperatures and include the production by electrolysis of metals such as aluminium and magnesium from their fused salts.
2 Those carried out below about 70°C using aqueous solutions and include electrolytic extraction and refining as well as electroplating.

In *extraction* processes (both in fused salts and aqueous electrolytes), an *inert* anode is used, whereas in *refining* processes a *soluble* anode is used.

The reactions which take place at the electrodes are as follows:

Cathode reaction   In all cases this can be represented as

$$M^{z+} + ze \rightarrow M$$

where z is a whole number.

Anode reaction   This will depend on the electrolyte used and also whether the anode is soluble or inert.

In **refining** processes, the anode (soluble) consists of the impure metal to be refined and metal ions are reformed by a reaction of the type:

$$M \rightarrow M^{z+} + ze$$

The electrolyte consists of an aqueous solution of a salt of the metal being refined.

In **extraction** processes the reaction at the anode (inert) depends on the electrolyte used, e.g.

Aqueous solutions	$2(OH)^- \rightarrow H_2O + \frac{1}{2}O_2 + 2e$
Fused chloride	$Cl^- \rightarrow \frac{1}{2}Cl_2 + e$
Fused oxide	$O^{2-} \rightarrow \frac{1}{2}O_2 + 2e$

2.4 Physical Chemistry of Extraction and Refining

Physical chemistry is concerned with the laws of chemical reaction, the extent and rate at which reactions take place, and with defining quantitatively the controlling factors such as temperature, pressure and concentration.

Energy Changes

Energy is the ability to do work and may manifest itself in several ways including kinetic, potential, electrical and thermal energy. Also, chemical compounds possess chemical energy which may be changed into heat. Chemical reactions are usually accompanied by either the evolution or the absorption of heat. Reactions in which heat is lost to the surroundings are said to be *exothermic*, while if heat is taken in the reaction is called *endothermic*. The heat change involved in a reaction may depend on the existing conditions,

Fig. 2.10 Energy change in a reaction

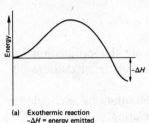

(a) Exothermic reaction
   $-\Delta H$ = energy emitted

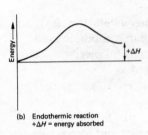

(b) Endothermic reaction
   $+\Delta H$ = energy absorbed

e.g. whether constant pressure or constant volume is being maintained. This effect is relevant only for those reactions in which there is a change in the number of moles of gaseous substances. When a reaction is carried out at constant volume, no work can be obtained, so that the heat change is the difference between the chemical energy of the reactant and the product molecules. Heat energy is evolved when the chemical energy of the reactants is greater than that of the products. The energy of reactants and products is referred to as *enthalpy* or *heat content*.

Reactions are usually carried out at constant pressure, namely that of the atmosphere. A heat change which accompanies a reaction performed at constant pressure is called an enthalpy change, and given the symbol ΔH. Enthalpy changes can be represented as shown in fig. 2.10. In an exothermic reaction, the enthalpy of the products is lower than that of the reactants, so that the enthalpy change, taken by convention to be,

ΔH = Enthalpy of products − Enthalpy of reactants

carries a negative sign.

For endothermic reactions, ΔH carries a positive sign.

The enthalpy change for a chemical reaction depends on:

a) The quantities of the reactants; the enthalpy change is conveniently written beside the balanced chemical equation, e.g.

$$CO + \tfrac{1}{2}O_2 \rightarrow CO_2 \qquad \Delta H = -283\,\text{kJ mol}^{-1}$$

b) The physical states of the reactants and products, e.g.

$$H_2 + \tfrac{1}{2}O_2 \rightarrow H_2O \text{ (liquid)} \qquad \Delta H = -284\,\text{kJ mol}^{-1}$$
$$H_2 + \tfrac{1}{2}O_2 \rightarrow H_2O \text{ (gaseous)} \qquad \Delta H = -242\,\text{kJ mol}^{-1}$$

c) The temperature at which the reaction is carried out. A standard temperature of 25°C is generally adopted, i.e. the initial temperature of the reactants and the final temperature of the products is 25°C (298K). When an enthalpy change relates to the standard temperature it is indicated as ΔH_{298K}.

If the initial and final temperature in a reaction is 298K, and the initial and final pressure one atmosphere, the enthalpy change is referred to as the *standard enthalpy change*, indicated by $\Delta H^{\ominus}$ (delta H nought).

Entropy

Most spontaneous reactions (i.e. reactions which are capable of proceeding of their own accord) are *exothermic* and result in a *decrease* in the enthalpy of the system. However, this is not the only determining factor and, in fact, some spontaneous reactions take place *endothermically*, so that heat is absorbed as they proceed, and there is a net *gain* in enthalpy. For example,

$$C(\text{solid}) + H_2O(\text{gas}) \rightarrow CO(\text{gas}) + H_2(\text{gas}) \qquad \Delta H = +131\,\text{kJ mol}^{-1}$$

The explanation is that another factor plays an important part in whether a reaction is spontaneous or not, and this factor is known as the *entropy* of the system. In simple terms, entropy may be regarded as a measure of the degree of disorder or randomness of the system. For example, a crystalline substance with a highly ordered structure usually has a low entropy, while the gaseous state has a high entropy. Changes of state are accompanied by entropy changes. Gases are more disordered than liquids, which in turn are more disordered than solids. Therefore, the changes, solid to liquid, liquid to gas, and solid to gas, are all accompanied by an *increase* in entropy (or disorder). Since entropy increases, the entropy change ΔS is positive.

It is convenient to regard a substance in the crystalline state at absolute zero temperature as having zero entropy because its atoms or ions are then in rigid order and perfect regularity. A change in entropy, ΔS, is expressed in JK^{-1}, and its value at 25°C and 1 atmosphere pressure is symbolised by ΔS^{θ}.

A change taking place of its own accord in an isolated system (i.e. one in which energy can neither enter nor leave) is always accompanied by an *increase* in entropy.

The enthalpy change ΔH represents the total amount of energy that accompanies a chemical or physical change. Not all of this energy is available for the isolated system to do work. Some is retained by the system in increasing the entropy. The term $T\Delta S$ is equal to this retained energy, and the difference

$$\Delta H - T\Delta S$$

expresses the amount of energy which is free or available to do work, i.e. it is the *free energy* change ΔG.

$$\Delta G \ = \Delta H - T\Delta S$$
or $\quad \Delta G^{\theta} = \Delta H^{\theta} - T\Delta S^{\theta} \quad$ under standard conditions

where $\quad T \ $ = absolute temperature
$\quad \Delta G^{\theta}$ = standard free energy change
$\quad \Delta H^{\theta}$ = standard enthalpy change
$\quad \Delta S^{\theta}$ = standard entropy change.

It is clear, therefore, that a reaction will occur of its own accord, without the supply of external energy, only if there is a *decrease* in free energy, i.e. if ΔG is negative. The free energy change is a measure of the driving force of a reaction under a given set of circumstances.

If values of ΔG are plotted against temperature, a straight line would be obtained if ΔH and ΔS were constants. Although this is not strictly the case, their variations with temperature are so small as to allow a linear plot using average values of ΔH^{θ} and ΔS^{θ}.

Ellingham first proposed the use of ΔG–T charts, and they have found wide application in rapid evaluation of data for metallurgical reactions. They give useful preliminary information on the conditions necessary for effective use of a proposed reaction. Figure 2.11 shows the variation in ΔG^{θ} values with temperature for a number of oxidations by reaction of the metal with one mole of oxygen gas at 1 atmosphere pressure. Since data for the formation of the oxides of carbon and for water are also given, some other metallurgical reactions may be assessed. The diagram shows the following features:

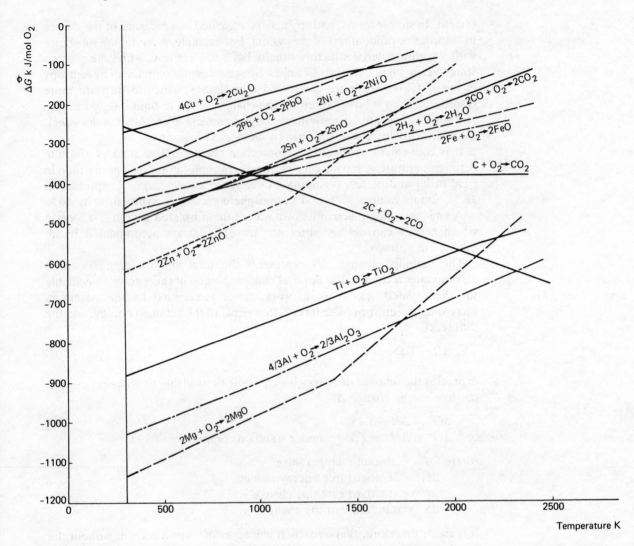

Fig. 2.11 A simplified Ellingham diagram for oxide formation

1 The slope of the lines is in general upwards from left to right, the free energy becoming less negative as the temperature increases. This is due to the decrease in entropy when a metal combines with oxygen to form a crystalline solid and to the disappearance of one mole of oxygen.

2 Some lines become steeper at certain points which correspond to the boiling points of the metals. This is again a result of entropy losses due to the disappearance of metal vapour, as well as oxygen gas when the oxidation reaction occurs.

3 The lower the position occupied by a metal oxide on the chart the more negative is the standard free energy, i.e. the more stable it is. Accordingly, at a given temperature and pressure, a metal is able to reduce the oxides of all other metals appearing above it on the chart. Therefore, it is possible to see at a glance whether a metal oxide is reducible by another metal (e.g. aluminium can be used to reduce iron oxide).

4 The slope of the carbon line is opposite to that of the other lines, due to the increase in the number of moles of gas $(2C + O_2 \rightarrow 2CO)$ and therefore an increase in the entropy. Hence the carbon lines intersects the other lines. The significance of the intersections is that they show the minimum temperature at which oxides can be reduced by smelting with carbon. The main reducing agents in smelting are carbon and carbon monoxide, so that the reactions

$$C + O_2 \rightarrow CO_2 \quad \text{and} \quad 2C + O_2 \rightarrow 2CO$$

are of great importance.

From the Ellingham diagram it is seen that the free energy change in the first reaction is practically independent of the temperature, and below about 710°C is greater than in the second reaction. In other words, CO_2 is more stable than CO below this point. The line for the second reaction, however, shows that the free energy of CO becomes increasingly negative with temperature and above 710°C, CO is more stable. The $C \rightarrow CO$ and $C \rightarrow CO_2$ lines intersect at 710°C and thus, at temperatures well below this point, the oxidation product of carbon during a reduction process will be CO_2. Further, consideration of the $C \rightarrow CO$, $C \rightarrow CO_2$ and $CO \rightarrow CO_2$ lines indicates that, below 710°C, CO is a more active reducing agent than carbon, but above it the reverse is true.

5 The reducing power of hydrogen does not increase significantly as the temperature is increased and therefore does not have the scope of carbon for reduction. Water gas (containing about 50% H_2 and 25% CO) may, however, be used for reducing some oxides (e.g. nickel oxide).

6 Traversing down the diagram, progressively more stable oxides are met, and the use of carbon as a reducing agent eventually becomes impracticable. Al, Mg and Ca then provide alternative choice as reducing agents.

In addition to the above diagram for oxides, further charts have been prepared for the formation of sulphides, chlorides and carbonates and are useful for studying extraction reactions. However, sulphides and chlorides are not usually suitable starting materials for metal extraction. The two most commonly used reducing agents, namely carbon and hydrogen, do not reduce sulphides because CS_2 and H_2S have less negative values for ΔG^θ than most metal sulphides. For example, in the case of copper sulphide,

$$2Cu_2S + C \rightarrow 4Cu + CS_2 \qquad \Delta G^\theta_{1100°C} = +160 \, \text{kJ} \, \text{mol}^{-1}$$
$$Cu_2S + H_2 \rightarrow 2Cu + H_2S \qquad \Delta G^\theta_{1100°C} = +130 \, \text{kJ} \, \text{mol}^{-1}$$

Sulphides are therefore converted into oxides which may be readily reduced, or alternatively into sulphates to be dealt with by hydrometallurgical methods.

Electrolytic Extraction from Ionic Melts

Electrolysis is very useful for extracting reactive metals from oxides whose free energy curves lie too far down the Ellingham diagram for reduction by carbon to be feasible at ordinary smelting temperatures. If a chemical reaction has an unfavourable free energy value, it may still be made to occur, provided enough electrical work can be brought to bear on it.

Quite small voltages are equivalent to large free energy changes but, in practice, rather high voltages are needed to overcome various losses. Reactive metals cannot be extracted from aqueous solutions because they displace hydrogen from solution. Ionic melts (e.g. chlorides and fluorides) which are molten below about 1000°C are therefore used as electrolytes. The lower melting point metals (e.g. Al, Mg) are the most suitable for extraction from fused salts, because they can be collected as liquid metal at the cathode and removed from the electrolytic cell at intervals. More refractory metals (e.g. Ti) are less suitable because they are more difficult to separate from the electrolyte and also may oxidise rapidly when removed.

Limitations of Free Energy Diagrams

Despite the fact that these diagrams provide very useful information, their application is limited for several reasons, including the following:

a) The diagrams refer to standard conditions but, in practice, this may not be the case. Non-standard conditions may result in wide deviation from diagram deductions.

b) The diagrams do not provide information regarding the rate of reactions. A reaction may be feasible but its rate so slow that it is of no value.

c) The diagrams do not indicate the conditions under which reactions are likely to occur.

In practice, concentration, temperature, pressure, particle size and catalysis may all be important.

2.5 Extraction of Iron

Ores

A common method of classifying iron ores is according to their appearance, together with their iron content, as shown in table 2.3.

Table 2.3 Classification of iron ores

Colour	Typical Fe content	Type	Name
Black	65%	Oxide (Fe_3O_4)	Magnetite
Red	50%	Oxide (Fe_2O_3)	Haematite
Brown	30–50%	Hydrated oxide $2Fe_2O_3.3H_2O$	Limonite
Yellowish brown	30%	Carbonate $FeCO_3$	Siderite

Iron ores can also be classified according to the type of gangue present, e.g. whether the gangue is rich in silica SiO_2 or rich in lime CaO. The main characteristics looked for in an iron ore are:

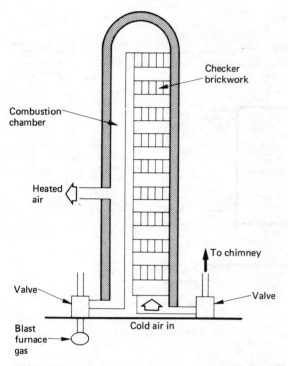

Fig. 2.13 Hot blast stove

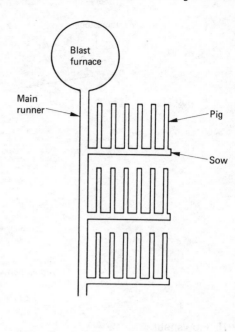

Fig. 2.14 Layout of pig beds

a) The combustion chamber in which cleaned blast furnace gas is burned.

b) A "honeycomb" of firebrick (known as "checkers") which provides a large surface area for both absorbing and giving up heat.

The cold air and the hot burned gas flow alternately in opposite directions through the stove, with the hot gas giving up its heat to the brickwork, and then the brickwork releasing this heat to preheat the cold air-blast. This method of heat exchange is known as *regeneration*. To enable a continuous supply of hot blast, each furnace has several stoves associated with it working in sequence.

Tapping Arrangements

Tapping of the iron from the furnace occurs about every two hours, through the iron tap hole. The clay which seals the tap hole is drilled out and, on completion of tapping, the hole is plugged using a "clay-gun".

The oldest method of casting iron (fig. 2.14) was to run it into beds, which had been prepared by moulding sand around patterns. The molten iron entered the bed by the main runner, while channels at right angles to this (known as "sows") fed small channels called "pig" beds—hence the name *pig iron*. The modern method is to use a pig casting machine, which consists of an endless chain belt carrying a series of cast iron moulds into which molten iron is delivered from a ladle. The product is free from adhering sand, and is of a more uniform composition for each cast.

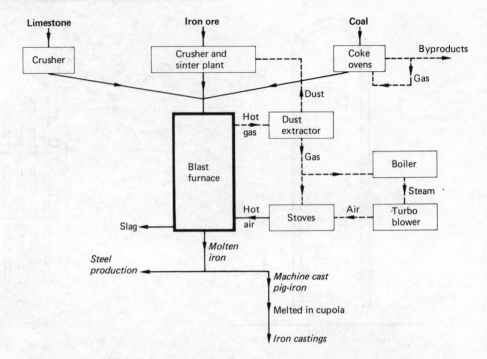

Fig. 2.15 Flowsheet for iron production

In an integrated iron and steel works, however, the molten iron from the blast furnace (known as hot metal) is taken directly to the steel plant, and poured into a holding furnace called a "mixer", from which molten iron is taken, as required, to the steelmaking furnaces.

The slag from the blast furnace is tapped frequently or even continuously through the slag notch, and it flows through runners (in the same way as the iron) into slag ladles or to a slag yard.

The gas led off from the top of the blast furnace, although it is a lean fuel, containing only about 30% CO, is useful for heating purposes. As it leaves the blast furnace, the gas contains a considerable amount of dust and grit, so that it must be cleaned before it can be used as a heating fuel. Cleaning is achieved by passing the gas through a dry dust catcher, and by using electrostatic dust catchers or wet scrubbers.

A simplified flowsheet for iron production is shown in fig. 2.15.

Chemistry of Iron Making in the Blast Furnace

The main reaction zones in the blast furnace are shown in fig. 2.16. Iron smelting in the blast furnace can be summarised as:

Iron ore + Coke + Limestone + Air → Iron + Slag + Gas

The reduction of iron oxide to metallic iron requires

a) A supply of heat.
b) The presence of a reducing agent.

Both of these requirements are met by the coke.

Fig. 2.16 Main reaction zones in the iron blast furnace

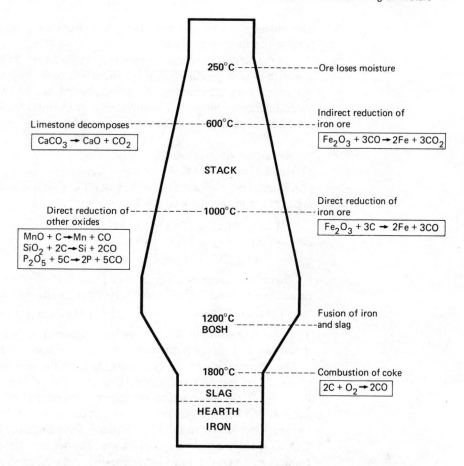

250°C — — — — — — — — — Ore loses moisture

Limestone decomposes — — — — — 600°C — — — — — — —

$$CaCO_3 \rightarrow CaO + CO_2$$

Indirect reduction of iron ore

$$Fe_2O_3 + 3CO \rightarrow 2Fe + 3CO_2$$

STACK

Direct reduction of — — — — — 1000°C — — — — — — — other oxides

$$MnO + C \rightarrow Mn + CO$$
$$SiO_2 + 2C \rightarrow Si + 2CO$$
$$P_2O_5 + 5C \rightarrow 2P + 5CO$$

Direct reduction of iron ore

$$Fe_2O_3 + 3C \rightarrow 2Fe + 3CO$$

1200°C — — — — — — — Fusion of iron
BOSH and slag

1800°C — — — — — — — Combustion of coke

$$2C + O_2 \rightarrow 2CO$$

SLAG

HEARTH
IRON

Three essential chemical processes take place in the furnace:

1   Combustion of the carbon in the coke with oxygen in the air blast.
2   Reduction of the iron oxide to iron.
3   Fluxing of the gangue of the ore, and of the ash from the coke, by the limestone.

The heat given out by the burning of the coke allows the reduction and fluxing processes to proceed.

1   *Combustion of the coke by the hot air blast*
The hot air blown through the tuyeres reacts with the coke to form CO, the overall reaction being

$$2C + O_2 \rightarrow 2CO \qquad \Delta H = -220 \, \text{kJ mol}^{-1}$$

Within the tuyere zone CO_2 is formed, but this reacts with coke outside the tuyere zone to give CO.

$$C + O_2 \rightarrow CO_2 \qquad \Delta H = -395 \, \text{kJ mol}^{-1}$$
$$CO_2 + C \rightarrow 2CO \qquad \Delta H = +175 \, \text{kJ mol}^{-1}$$

The burning of the coke is a strongly exothermic reaction, and the temperature at the tuyeres is about 1800°C.

Iron making in the blast furnace is a counter-current process so that the hot gas, rich in CO, produced in the combustion zone, rises up the stack, and heats and reacts with the charge descending towards the hearth. The nitrogen in the ascending gases, although it plays no part chemically, acts as a heat transfer medium helping to heat the descending charge.

2 *Reduction of iron oxides*

a) Reduction by CO, i.e. *indirect reduction*, starts at about 300°C and predominates up to about 1100°C when the slag starts to form. The reduction takes place in the furnace stack in stages:

$$3Fe_2O_3 + CO \rightarrow 2Fe_3O_4 + CO_2$$
$$Fe_3O_4 + CO \rightarrow 3FeO + CO_2$$
$$FeO + CO \rightarrow Fe + CO_2$$

b) *Direct reduction* by carbon takes place further down the stack at temperatures above about 1000°C. and is typified by the equation

$$FeO + C \rightarrow Fe + CO \qquad \Delta H = +156\,kJ\,mol^{-1}$$

The above reaction being endothermic is favoured by a high temperature and it follows that direct reduction occurs in the hotter furnace zones.

The composition of the gas emerging from the furnace depends on the ratio of direct to indirect reduction. For example, if the conditions lead to an increase in the proportion of direct reduction then the CO/CO_2 ratio is increased. The CO/CO_2 ratio is used as one of the main control parameters in the smelting of iron in the blast furnace. Economically it is important that the CO/CO_2 ratio should be kept as low as possible consistent with satisfactory reduction, since the ratio influences the coke used in making 1 tonne of pig iron (i.e. the coke rate). Direct reduction of other oxides in the burden also occurs and this takes place mainly in the upper bosh. For example, SiO_2 and MnO are reduced by carbon to silicon and manganese respectively, which then dissolve in the descending iron.

In gas-solid reactions the rates of reaction are governed by the amount of surface area of solid presented to the gas. Therefore a porous mass (e.g. agglomerate) is more easily reduced and favours indirect reduction.

Reduction to iron is complete well above the bosh, i.e. before the iron is melted. This solid product is called sponge iron, and is fairly free from impurities, but as it descends the furnace it soon dissolves carbon and other elements.

The gas produced in the combustion zone rises rapidly inside the furnace, and the concentration of CO in the upper parts of the stack is higher than would be expected. In the temperature range 550–650°C finely divided carbon forms as a result of the reaction

$$2CO \rightarrow C + CO_2$$

This carbon may cause a small amount of direct reduction of iron oxide to take place in the upper stack.

3 *Fluxing by limestone*

The main constituents of the gangue from the ore, and the ash from the coke, are oxides with very high melting points (e.g. SiO_2 and Al_2O_3) and these could not be melted separately at blast furnace temperatures. Therefore limestone is added to the furnace charge to act as a flux.

At about 600°C the limestone ($CaCO_3$) is calcined to CaO:

$$CaCO_3 \rightarrow CaO + CO_2 \qquad \Delta H = +175 \, kJ \, mol^{-1}$$

Because this calcination is an endothermic reaction it increases the thermal load on the furnace, so that if calcination is carried out during sintering then the coke-rate of the furnace would be reduced. Also the presence of less CO_2 inside the blast furnace increases the reducing potential of the gases in the furnace.

The liquid slag when first formed in the upper bosh is rich in FeO and consists mainly of FeO, SiO_2 and Al_2O_3. As it descends, CaO enters the slag and its basicity increases. At hearth level, more SiO_2 and Al_2O_3 from the coke-ash enter the slag, leading to a decrease in the basicity. A typical composition of blast furnace slag would be 35% SiO_2, 40% CaO, 15% Al_2O_3, 5% MgO. The composition of the slag has a marked influence on the changes which take place in the composition of the iron in the hearth.

Apart from the SiO_2 which enters the slag, some SiO_2 is reduced by carbon at high temperatures (as mentioned above) and Si enters the iron. Raising the temperature results in increased reduction of SiO_2 to Si which dissolves in the iron. This reaction is

$$SiO_2 + 2C \rightarrow Si + 2CO$$

However, increasing the basicity of the slag favours the entry of SiO_2 into the slag and thus lowers the pick-up of Si into the iron.

In the case of manganese, raising the hearth temperature and increasing the slag basicity both favour the reaction

$$MnO + C \rightarrow Mn + CO$$

resulting in higher Mn content in the iron.

The reduction of P_2O_5 also occurs by reaction with carbon, but the phosphorus formed enters the iron before the onset of slag formation and nearly all the phosphorus present in the burden ends up in the iron:

$$P_2O_5 + 5C \rightarrow 2P + 5CO$$

The sulphur in the burden is present mainly in the coke, and its partial removal into the slag depends on the formation of a more stable sulphide than FeS (namely CaS) and needs a basic slag rich in CaO, as well as a high hearth temperature:

$$FeS \, (in \, iron) + CaO \rightarrow CaS \, (in \, slag) + FeO$$

In the hearth of the furnace there is opportunity for the various elements to partition between the iron and slag, and this depends largely on the hearth temperature and the composition of the slag (i.e. its basicity). To ensure that the slag is sufficiently fluid, temperatures in the range 1400–1500°C are necessary.

Summarizing, the composition of the iron produced depends on the following inter-related factors:

1) Burden composition   2) Slag composition   3) Hearth temperature

Efficient blast furnace operation, i.e. achieving the maximum amount of indirect reduction and operating the process at the lowest temperature to give the required production rate, demands careful burden preparation, including correct ore blending and use of agglomerate with the right texture.

Significant improvements in furnace efficiency and production rate have resulted from making various additions (e.g. oxygen, hydrocarbons, steam) to the air blast, as well as increasing the gas pressure at the top of the stack, thus allowing the mass of air through the furnace to be increased without increasing the air velocity. However, these practices have to be carefully controlled to avoid steep temperature gradients resulting in irregular descent of the charge (a "hanging" burden).

Approximate relative proportions by weight of raw materials required to produce 1 tonne of iron are given in table 2.4.

Table 2.4 Raw material proportions required for 1 tonne of iron

Raw Materials (tonnes)		Products (tonnes)	
Iron ore (50% Fe)	2	Iron	1
Coke	0.4–0.6	Slag	0.3
Limestone	0.3	Gases	5
Air	2.5 ($3 \times 10^6 \, m^3$)		

The carbon content of blast furnace iron varies from about 3.5 to 4.5%, while the sulphur content is usually below about 0.05%, and the amount of manganese generally lies between 0.4 and 1%. Large variations in the phosphorus content are possible, depending on the type of iron ore used, while variation in the operating conditions of the furnace can result in a significant difference in the silicon content of the iron produced. Some typical iron *compositions* are given in table 2.5.

Table 2.5 Composition of various types of iron

Type of iron	P/Si content	Uses
Haematite iron	Low P 0.04% max. High Si 2–3%	High duty iron castings
Foundry iron	High P 0.7–1.2% High Si 2–3%	General foundry work
Iron for steelmaking a) Medium phosphorus b) High phosphorus	P up to about 0.4% P up to about 2%	Steelmaking by Basic Oxygen Process

Disadvantages of blast furnace smelting

1) High capital and operating costs are involved.
2) Control of iron composition is rather poor.

3) Small coke-using furnaces are inefficient, while huge outputs from large furnaces may not be required.

Alternatives to blast furnace operation include the following:

1 The use of a small-stack furnace in which cheaper carbonaceous materials (e.g. coal, lignite) can achieve direct reduction. The lower production rate is convenient for meeting smaller demand.

2 Electric arc smelting where low-cost electricity is available. Reduction is again mainly by direct reaction with carbon.

3 The production of sponge iron from pelletised iron ore, without going through the molten stage, which is used as a feedstock for electric arc steelmaking. The ore pellets may be reduced by solid carbonaceous material (e.g. coal dust or coke breeze) in rotary kilns or alternatively by a gaseous reducing agent in a low stack furnace. The processes are carried out at about 1000°C. The gaseous reducing agent consists of a mixture of CO and hydrogen which enters the shaft furnace where it meets the descending iron ore in a counter current flow. The reduction may be represented by the equations

$$Fe_2O_3 + 3H_2 \rightarrow 2Fe + 3H_2O$$
$$Fe_2O_3 + 3CO \rightarrow 2Fe + 3CO_2$$

The iron produced passes through a cooling zone and is then discharged at about 25°C.

The control of the composition of the iron is better than that attainable in the blast furnace.

2.6 Steelmaking

The principle involved in converting blast furnace iron into steel is that certain impurities present in the molten iron are selectively oxidised and removed from the melt. The impurities removed in this way include carbon, silicon, manganese and phosphorus, and their amounts are decreased to suitable levels to meet the required specification. The removal of sulphur is exceptional in that oxidation is not involved and, in fact, desulphurisation is most efficient in de-oxidising conditions.

After removal of impurities, it is necessary to remove the excess oxygen remaining in the metal at the end of the refining. De-oxidation is achieved by adding elements which have a greater affinity than iron for oxygen. These added elements form oxides which are insoluble in iron and are removed into the slag.

The Basic Oxygen Steelmaking Process

In 1952 in the Austrian towns of Linz and Donawitz, a steelmaking process was established which used oxygen gas for the oxidation of impurities in iron and was known as the LD process. This process has become the most important method of making steel, and is now known as the BOS process.

The steelmaking vessel (fig. 2.17) is tilted and charged first with steel scrap

Fig. 2.17 BOS
converter

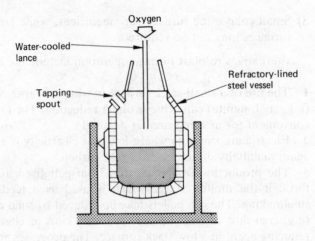

(about 30% of the charge), then with molten blast furnace iron, and returned to the upright position. A water-cooled oxygen lance is then lowered to within about a metre of the molten iron and high purity oxygen is blown on to the iron. During the early stages of the blow, lime (CaO) is added and this combines with oxidised impurities to form a slag. It is important that a good, fluid, basic slag is formed early in the process because this speeds up removal of impurities from the iron. During the blow, the impurities are removed in the following order: silicon, manganese, phosphorus, carbon and sulphur. Carbon oxidises to form CO which is insoluble in the slag and escapes to the mouth of the converter where it burns to form CO_2.

When the blow is completed, the composition is adjusted by additions to the melt and the steel is poured into a ladle. Afterwards, the converter is turned upside down and the slag is tipped into a slag ladle. The steel in the ladle is de-oxidised by adding aluminium, silicon (added as an alloy with iron, i.e. ferro-silicon), or manganese (added as ferro-manganese).

Modern BOS converters have capacities of about 300 tonnes, and the charge can be converted into steel with a blowing time of about 20 minutes and a tap-to-tap time of about 40 minutes.

Chemistry of the BOS Process

The removal of impurities as blowing proceeds is represented in fig. 2.18. During the blowing there is some reaction between molecular oxygen and the impurities, while iron is also oxidised to iron oxide, which dissociates to provide atomic oxygen to the melt at the slag-metal interface. The supply of oxygen is, therefore, controlled by the fluidity of the slag, turbulence, and area of contact of slag and metal. The oxidation reactions may be expressed in simple terms by the following equations:

$$2Fe + O_2 \rightarrow 2FeO$$
$$FeO + Mn \rightarrow Fe + MnO \qquad 5FeO + 2P \rightarrow P_2O_5$$
$$2FeO + Si \rightarrow 2Fe + SiO_2 \qquad FeO + C \rightarrow Fe + CO$$

Fig. 2.18 Removal of impurities from low P iron in BOS process

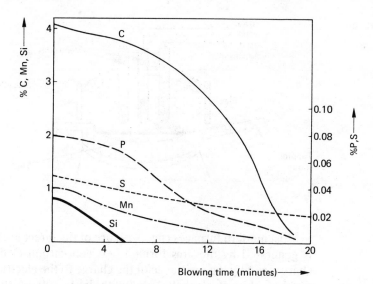

The above reactions, involving oxidation of the impurities, are strongly exothermic and provide the heat required to raise the temperature of the melt as refining proceeds. The P_2O_5 formed combines with the added lime to form calcium phosphate which enters the slag:

$$P_2O_5 + 3CaO \rightarrow Ca_3(PO_4)_2$$

Similarly, MnO and SiO_2 are fluxed and taken into the slag.

The removal of sulphur demands a slag rich in CaO, a high melt temperature, and melt conditions which are not strongly oxidising. It follows that only limited removal of sulphur can be achieved in the BOS process because the oxygen potential of the metal is rather high. The reaction may be represented by

$$FeS + CaO \rightarrow CaS + FeO$$

Electric Arc Steelmaking

Electric arc furnace melting of metals enables high temperatures to be quickly attained, so that a rapid rate of melting can be achieved. Further, close control of temperature and the conditions inside the furnace are possible, and this allows low-grade scrap to be used to produce high-quality steel with very low sulphur content. In fact, the charge may consist entirely of steel scrap with no pig iron at all. The process then simply involves the re-cycling of steel scrap.

Arc furnaces are used for making high-quality carbon and alloy steels and have a capacity up to about 120 tonnes. The use of this type of furnace is increasing rapidly, especially for the production of steel from 100% scrap charges.

Fig. 2.19 Electric arc furnace

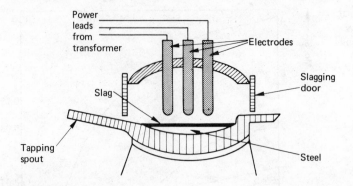

Most modern furnaces (fig. 2.19) are of the direct arc type in which the heat is generated by the arcs formed between carbon electrodes and the metal charge, and by the resistance of the charge to the electric current. The furnace consists of a steel shell lined with basic refractory bricks. The roof is dome-shaped and has three openings through which the carbon electrodes can move freely. These electrodes burn away and new pieces are added as their length shortens. The furnace can be tilted forward for tapping the steel or backward for removing the slag.

The electric arc steelmaking process may be divided into four stages as follows:

Charging   The roof is swung aside and the charge, consisting of steel scrap, pig iron, iron ore (iron oxide) and lime are placed in the furnace. The roof is swung into place, the electrodes lowered, and the current switched on.

Melting   The charge starts to melt and the silicon, manganese and phosphorus present begin to be oxidised to their respective oxides and combine with the lime to form a basic slag. The oxidation reactions taking place are similar to those stated previously for the BOS process.

Refining   The removal of phosphorus demands a basic slag rich in CaO, a low metal temperature, and strongly oxidising conditions in the melt (i.e. high oxygen potential in the metal). After the charge has become completely molten, an oxygen injection lance may be used to speed up the oxidation. The removal of carbon now takes place and CO is formed, which bubbles through the melt, giving rise to turbulence, known as "the carbon boil".

The slag at this stage is known as *black oxidising slag*, and must be run off, otherwise the phosphorus would return to the molten metal during the next stage. The black slag is immediately replaced by a *white reducing slag*, which is made by adding lime, fluorspar (calcium fluoride CaF_2) and anthracite coal. The white, basic slag provides the reducing conditions (i.e. metal with low oxygen potential) which enables good removal of sulphur to be achieved. The sulphur enters the slag as calcium sulphide. The reaction may be represented by:

$$FeS + CaO + C \rightarrow CaS + Fe + CO$$

Finishing   The molten metal is brought to the required composition by the addition of alloys and the temperature is adjusted to that required. The electrodes are raised and the furnace tilted to pour the steel into a ladle, the slag being held back until the end of the pour. The steel is then deoxidised using aluminium, ferro-silicon or ferro-manganese.

The electric process is slower (120 tonnes in 4 hours) than the BOS process. However, it allows much greater control over the refining reactions and achieves excellent removal of sulphur compared with the BOS process. Arc furnace charges may consist of 100% scrap or contain up to 50% molten iron. The electric arc process is widely used for producing high-quality carbon steel and alloy steels.

2.7 Extraction of Aluminium

Although aluminium is the most abundant metal in the earth's crust, much of it is present as aluminium silicates in various clays, from which it is very difficult to economically extract the aluminium. However, deposits of hydrated oxide exist from which aluminium can be profitably recovered.

Ore

The ore is called *bauxite*, which is impure hydrated alumina (mainly $Al_2O_3.3H_2O$). The main impurities are ferric oxide Fe_2O_3 (which gives a brown colour to the bauxite), silica SiO_2, and titanium dioxide TiO_2. Bauxite usually contains 45–60% Al_2O_3, 1–12% SiO_2, 2–25% Fe_2O_3, 15–35% combined water, and up to about 5% TiO_2. Top-grade bauxite is in limited supply. Some of the important sources of bauxite are Australia, Jamaica, Ghana, Guyana and France.

Ore Preparation

The bauxite is crushed and calcined at the mine (to destroy organic matter and to remove moisture) then transported to the aluminium reduction plant.

Extraction

Aluminium is a very reactive metal and forms an extremely stable oxide Al_2O_3. Consequently, reduction of the oxide by carbon or CO would require very high temperatures (above 2000°C, see fig. 2.11) and is not a practical proposition. The high stability of Al_2O_3 also means that conventional methods of refining the metal by selective oxidation of impurities would be impractical. It follows that the impurities in bauxite must be removed before reduction of the oxide to metal.

In practice the extraction of aluminium consists of two main stages:

1   The preparation of pure Al_2O_3 from bauxite by the Bayer process.
2   The electrolysis of a fused salt mixture containing Al_2O_3.

The **Bayer process** for purification of bauxite involves four stages as illustrated in fig. 2.20.

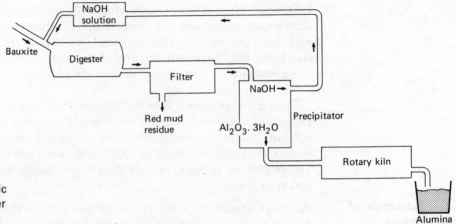

Fig. 2.20 Diagrammatic representation of Bayer process

a) **Digestion**   The crushed and calcined bauxite is dissolved in a hot (150°C), strong (30%) solution of caustic soda, at a pressure of about 6 atmospheres, in reaction vessels called *autoclaves*. The Fe_2O_3 and TiO_2 impurities remain undissolved, but some of the SiO_2 forms sodium aluminium silicate, and this represents a loss of Al_2O_3 and NaOH to the process. The Al_2O_3 dissolves to form sodium aluminate solution:

$$Al_2O_3.3H_2O + 2NaOH \rightarrow Na_2O.Al_2O_3 + 4H_2O$$

b) **Filtration**   The solution is diluted and the insoluble "red mud" residue allowed to settle in slowly-stirred settling tanks. The solution is then filtered and the red mud (containing Fe_2O_3, SiO_2 and TiO_2) discarded.

c) **Precipitation**   The sodium aluminate filtrate is pumped to large tanks, in which the solution is "seeded" with very small particles of freshly precipitated aluminium hydroxide. Compressed air is blown into the tanks, and this agitation, together with slow cooling, causes the decomposition of the sodium aluminate to produce $Al_2O_3.3H_2O$.

$$Na_2O.Al_2O_3 + 4H_2O \rightarrow Al_2O_3.3H_2O + 2NaOH$$

The sodium hydroxide solution regenerated is recycled to the start of the process.

d) **Calcination**   After about 36 hours, the precipitated $Al_2O_3.3H_2O$ is removed by vacuum filtration and transferred to rotary kilns, in which calcination is carried out about 1100°C to remove the chemically combined water and to produce αAl_2O_3. The very high calcining temperature is necessary to ensure that the Al_2O_3 is in the α form and not the γ form, which absorbs water vapour and would carry this moisture into the next stage (electrolysis) of the extraction, where reaction with the fluoride electrolyte would produce hydrofluoric acid.

The resulting calcine is a white powder and is 99.5% pure Al_2O_3.

Electrolytic Reduction of Aluminium Oxide

The dissociation of Al_2O_3 into aluminium and oxygen may be achieved by electrolysis, and theory shows that a relatively low voltage is required. Because aluminium is more electronegative than hydrogen, electrolysis of an aqueous solution containing aluminium would evolve hydrogen at the cathode and would not produce a deposit of Al. However, Hall (in U.S.A.) and Héroult (in France) discovered that molten cryolite Na_3AlF_6 was able to dissolve Al_2O_3, and that the solution could be electrolysed to produce aluminium, without appreciably decomposing the cryolite, though a very small amount of fluorine is evolved.

Cryolite melts at about 1000°C and the addition of about 5% alumina (from the Bayer process) together with a small addition of calcium fluoride lowers this enough for electrolysis to be carried out at about 900°C. An electrolyte consisting of a fused salt mixture is necessary because the melting point of Al_2O_3 (2020°C) is far too high for it to be used alone.

Fig. 2.21 Diagram of aluminium reduction cell

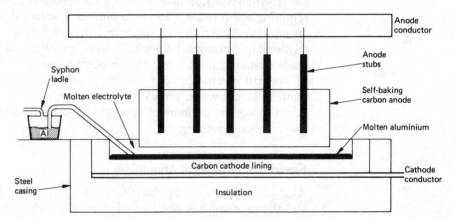

The electrolysis is carried out in cells known as "pots" (fig. 2.21). Blocks of carbon, formed by baking a mixture of petroleum coke and pitch, serve as the anode, through which the electric current enters the pot. The anodes dip into the electrolyte and are consumed by oxidation to CO_2. Renewal of anodes is either by replacing with pre-baked blocks or, alternatively, self-baking anodes are used, which involves the feeding of carbon paste into a steel tube above the electrolyte—the paste being hardened as it descends towards the pot. Under the anode is a large rectangular steel box lined with carbon; this lining is the cathode through which the current leaves the pot.

During electrolysis the aluminium produced collects in the bottom of the pot at the cathode, and the oxygen evolved at the anode reacts with the carbon to form CO_2. Any impurities present which are more noble than Al (e.g. Fe, Si) will also deposit at the cathode. The aluminium produced is about 99.5% pure and periodically it is syphoned off (using a vacuum ladle which sucks up the molten metal) or tapped from the bottom into crucibles. The metal is then cast into ingots, which vary in shape depending on the type of subsequent processing, and range from about 2.5 kg to about 500 kg.

The reactions taking place may be expressed in simple terms as follows:

Dissociation of alumina $\qquad\qquad Al_2O_3 \rightarrow 2Al^{3+} + 3O^{2-}$

Reaction at cathode $\qquad\qquad 2Al^{3+} + 6e \rightarrow 2Al$

Reaction at anode $\qquad\qquad\quad 3O^{2-} - 6e \rightarrow \frac{3}{2}O_2$

$\qquad\qquad\qquad\qquad\qquad\qquad C + O_2 \rightarrow CO_2$

Overall reaction $\qquad\qquad\quad 2Al_2O_3 + 3C \rightarrow 4Al + 3CO_2$

The mixture of gases at the anode, including a small amount of fluorine is collected and fed into a burner, the products of which are discharged from a high chimney. The amount of aluminium deposited is less than the theoretical quantity expected by application of Faraday's laws, giving an efficiency of 80–90%.

The electrolytic cells are connected in series and are operated continuously at about 6 volts. Since the molten bath has some resistance, heat will be generated and this keeps both the electrolyte and the aluminium molten. The electrolytic reduction of alumina to aluminium is a power intensive process, requiring about 18 kWh/kg of aluminium. The industry, therefore, represents an enormous total power requirement, and it is an economic necessity to locate aluminium smelters where large quantities of electric power can be made available at low-cost. Hydro-electric power used to be the best source of low-cost electricity, and was the foundation on which the industry was built. In recent years, however, and in countries where hydro-electric power is not available, aluminium smelters have been located near other energy sources, e.g. natural gas.

Electrolytic Refining of Aluminium

Aluminium is refined by electrolysis of a three-layer molten bath (fig. 2.22) and the procedure is known as Hoope's method. The bottom layer, which forms the anode, is the impure aluminium to which up to about 30% copper has been added to make the alloy denser so that it will stay at the bottom. On top of this is a layer of molten electrolyte, whose melting point must be above that of pure aluminium, and whose density must be higher than that of pure aluminium, but less than that of the bottom layer. The electrolyte consists of a mixture of aluminium fluoride and sodium fluoride together with barium fluoride and barium chloride to provide the required density.

Electrolysis is carried out at 700–900°C, during which aluminium dissolves from the impure anode and passes through the electrolyte (where it is further purified) and deposits on a carbon electrode at the top of the cell. The impurities and alloying elements in the anode, being more noble than aluminium do not dissolve. As the amount of impurity in the bottom layer increases it is withdrawn and replaced by new metal to be refined. The purity of the refined aluminium is 99.99+%. The capital costs of the refining cells are high and the electrical energy requirement is about 20 kWh/kg of refined aluminium. Using a cell voltage of 5 volts a current efficiency of about 90% is achieved. It is clear, therefore, that the refining of aluminium is expensive, but it is an important process for the production of high purity aluminium, which is used as a corrosion—resistant lining in processing vessels.

Fig. 2.22 Electrolytic
cell for refining
aluminium

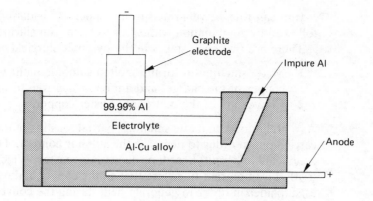

2.8 Pyrometal- lurgical Extraction of Copper from Sulphide ores

Ores

Copper is present in the earth's crust mainly in the form of sulphide minerals, often in conjunction with iron sulphide FeS, e.g.

Chalcocite Cu_2S (black)
Chalcopyrite $CuS.FeS$ (yellowish)
Bornite $Cu_2S.CuS.FeS$ (bluish)

The concentration of the above minerals in an ore-body is low, typical copper contents being about 0.5–2%.

About 90% of the world's primary copper comes from sulphide ores. The main impurity is iron, which may be accompanied by small amounts of Ni, Au and Ag and traces of Zn, Sn, Pb, Co, As, Sb, Se, Te and Bi. (Low-grade ores containing oxidised minerals also occur, but these are nearly always treated by hydro-metallurgical methods—see section 2.14.)

Some of the important sources of copper are, Zambia, Katanga (Congo), Chile, Canada (Sudbury), U.S.A. (Montana) and Australia (Mount Isa).

Ore Preparation

As mentioned above, the sulphide ores of copper are lean, with a copper content as low as 0.5%, and froth flotation is used to produce a concentrate containing 20–30% Cu, after crushing and grinding of the ore.

Extraction

At one time copper sulphide concentrate was completely roasted to copper oxide which was then reduced by carbon to the metal. This method was discontinued because the copper produced was very impure, and loss of metal to the slag was very high. The method now used depends on the following relationships at high temperature:

a) Copper has a greater affinity than iron for sulphur.
b) Iron has a greater affinity than copper for oxygen.

Iron can thus be separated from copper by oxidation to form iron oxide, followed by combination with silica to form iron silicate slag.

There are two main stages in the pyrometallurgical extraction of copper:

1 Matte smelting to form a molten sulphide melt, which contains all the copper of the charge, and a molten slag free from copper.
2 Conversion of the matte into blister copper.

As a preliminary to matte smelting, partial roasting of the concentrate may be carried out in order to decrease the sulphur content. This operation may be conducted in a multi-hearth or fluo-solids roaster. Whichever roasting unit is used, it is important to ensure that all the copper and part of the iron remain as sulphides in order to generate heat during the converting of the matte.

Matte Smelting

As stated in section 2.1 the reverberatory furnace (fig. 2.3) is being superseded by flash-furnace smelters (fig. 2.4) for matte smelting. The dried, fine concentrate (or calcine) is blown into the furnace, where it is combusted with oxygen or oxygen-enriched air. During the smelting some of the iron sulphide is oxidised to iron oxide and fluxed with silica to form iron silicate slag:

$$2FeS + 3O_2 \rightarrow 2FeO + 2SO_2 \qquad 2FeO + SiO_2 \rightarrow 2FeO.SiO_2$$

The rest of the iron remains as FeS, which mixes perfectly with copper sulphide to form a mixture of molten sulphides (matte). The slag is lighter than, and immiscible with, the matte and is tapped off separately. The matte, containing 25–50% Cu together with any precious metals (e.g. Ag, Au) present in the ore, is run off into a converter. The combustion reaction provides most of the heat required for heating and melting the matte and slag. The concentration of SO_2 in the effluent gases is high (usually more than 10%) and the SO_2 can be used to produce sulphuric acid, elemental sulphur or liquid SO_2.

Conversion of the Matte into Blister Copper

Converting consists of oxidising the molten matte with air or oxygen, resulting in the removal of iron and sulphur and the production of a crude "blister" copper containing about 98% Cu.

Molten matte is poured into a horizontal converter (fig. 2.5) and the air is introduced through the tuyeres. The heat generated by the oxidation of the FeS is sufficient to make the process autogenous. Converting is carried out in two stages:

a) The FeS elimination or slag-forming stage
Since FeO is more stable than Cu_2O, the FeS is oxidised in preference to Cu_2S and the FeO formed is fluxed with SiO_2 to form iron silicate slag. Any Cu_2O present will react with FeS as follows:

$$Cu_2O + FeS \rightarrow Cu_2S + FeO$$

The slag is poured off by tilting the converter and more matte is added: the process is repeated until the converter is nearly full of molten copper sulphide.

b) *The blister copper-forming stage*

When all the iron sulphide has been oxidised, copper production begins.

Firstly, Cu_2S is oxidised to Cu_2O:

$$2Cu_2S + 3O_2 \rightarrow 2Cu_2O + 2SO_2$$

then the reaction

$$Cu_2S + 2Cu_2O \rightarrow 6Cu + SO_2$$

takes place because it has a negative free energy value at the operating temperature (1200°C). The net reaction can be expressed as

$$Cu_2S + O_2 \rightarrow 2Cu + SO_2$$

The crude copper is about 98% pure, and contains 0.02–0.1% sulphur. If allowed to solidify, the metal is porous and brittle, and is known as *blister copper* due to the blisters on the surface caused by the evolution of SO_2.

Any precious metals present in the matte are not oxidised during conversion and enter the blister copper. The blister copper may then be purified by fire-refining.

Fire-refining of Blister Copper

This involves controlled oxidation of impurities followed by deoxidation of the copper.

Air is blown into the molten blister copper in a reverberatory furnace, and the more reactive impurities (e.g. Fe, Pb, Zn) are oxidised and removed to the slag. The addition of sodium carbonate, sodium nitrate and lime to the slag helps to remove other impurities such as As, Sb and Sn. The copper also starts to oxidise and air blowing is continued until the oxygen content of the copper reaches about 0.9% in order to ensure satisfactory removal of sulphur. However, copper having such a high oxygen content would be brittle so that deoxidation is required.

The slag is removed and the copper is stirred with tree trunks—an operation which is called "*poling*"—the hydrocarbons evolved from the green wood reducing the copper oxide. Poling is a critical operation in which the aim is to bring the oxygen content down to 0.03–0.06%. If the oxygen content is not lowered enough, the copper is mechanically weak, whilst if taken too low the solidified copper is porous due to the formation of steam pockets caused by the interaction of Cu_2O and reducing gases during solidification:

$$Cu_2O + H_2 \rightarrow 2Cu + H_2O$$

Poling is continued until the set ("pitch") of the surface of a test ingot is more or less horizontal and the copper is said to be in the "tough pitch" condition. Tough pitch copper contains 0.03–0.06% oxygen and is very suitable for working. If the copper is over-poled it must be re-oxidised and re-poled.

Final deoxidation of fire-refined copper may be achieved by adding

Fig. 2.23 Simplified
flowsheet for
pyrometallurgical
extraction of copper
from copper sulphide

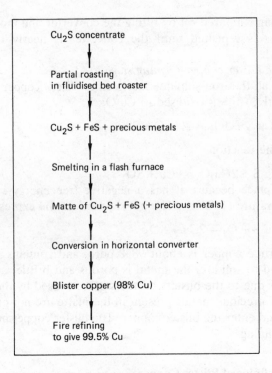

Fig. 2.23 Simplified flowsheet for pyrometallurgical extraction of copper from copper sulphide

phosphorus in the form of a Cu/14% P alloy. However, the small amount of phosphorus remaining in the Cu (0.05%) adversely affects the electrical conductivity and, to avoid this, lithium may be used as the deoxidant.

Fire-refining produces copper which is about 99.5% pure, and electro-refining is necessary to obtain higher purity.

A simplified flowsheet for the pyrometallurgical extraction of copper is shown in figure 2.23.

2.9 Pyrometallurgical Extraction of Lead and Zinc

Many attempts had been made to produce zinc by blast furnace smelting before the problems involved were finally solved by the Imperial Smelting Corporation at Avonmouth. The method developed involves the simultaneous reduction of a mixture of zinc oxide and lead oxide produced by sintering a flotation concentrate obtained from a sulphide ore.

Ores

The main ore of zinc is zinc blende or sphalerite ZnS, which is accompanied by considerable quantities of galena PbS. Chalcopyrite $CuS.FeS$ is also usually present, as well as small, but profitable, amounts of Ag_2S. Calcite ($CaCO_3$) and quartz (SiO_2) are common gangue minerals associated with zinc blende.

The main sources include U.S.A. (Missouri, Kansas), Canada (British Columbia, New Brunswick), Australia (Broken Hill, Mount Isa), and Mexico, which together produce about 60% of the world output of zinc.

Ore Preparation

The finely ground ore is treated by flotation to give a mixed ZnS/PbS concentrate. It is not necessary to separate the ZnS and PbS for blast furnace extraction.

The sulphides must be changed into oxides, and since the blast furnace requires granular material as feedstock the oxidation is most conveniently achieved by sintering. An up-draught machine is used, with a long hood in order to collect the SO_2 gas evolved, which is converted into H_2SO_4. It is most important not to allow the oxidation to develop too quickly, otherwise fusion will take place, thus preventing oxygen reaching the centre of the sulphide particles, so that some sulphide would remain. The sintering may be carried out in two stages and the charge for the second stage may consist of some pre-sintered material along with raw sulphide concentrate. By adjusting the speed of the endless belt on the machine, the process can be controlled to give the required degree of oxidation and agglomeration.

The exothermic, roasting reactions are

$$2ZnS + 3O_2 \rightarrow 2ZnO + 2SO_2$$
$$2PbS + 3O_2 \rightarrow 2PbO + 2SO_2$$

Extraction

The sinter mixture of ZnO and PbO is charged along with preheated coke and limestone at the top of a rectangular-section blast furnace, while air at a temperature of about 900°C is blown through tuyeres at the bottom of the stack (fig. 2.24).

The combustion of the coke produces CO ($2C + O_2 \rightarrow 2CO$) which reduces PbO at about 500°C in the top third of the stack:

$$PbO_s + CO_g \rightarrow Pb_l + CO_{2g}$$

This gas-solid reaction is exothermic and does not consume additional coke. Reduction of PbO by carbon occurs only slightly: this solid-solid reaction being much less favourable than reduction by CO.

The liquid lead collects in the hearth and is tapped from the bottom of the furnace. In addition to small amounts of silver, the lead may contain varying amounts of Cu, Sn, As, Sb, Bi and Fe. The lead may be refined by either pyrometallurgical methods or electrolytically.

The reduction of ZnO is more difficult because reaction with carbon takes place only above 950°C—which is above the boiling point of zinc (907°C):

$$ZnO_s + C_s \rightarrow Zn_g + CO_g$$

The reduction of ZnO by coke takes place in the hottest part of the stack (i.e. in the bottom third), and the CO produced in the reaction ascends the furnace to reduce PbO in the top third of the stack. The zinc produced is in vapour form and will be oxidised to ZnO by any CO_2 present if the temperature falls below 950°C because ZnO is more stable than CO_2 below this temperature. In order to maintain the temperature above the re-oxidation temperature, a controlled amount of preheated air is blown in at

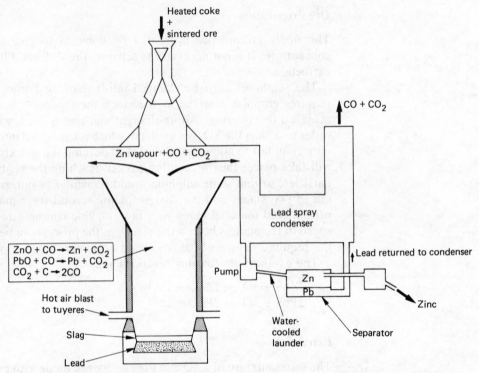

Fig. 2.24 Zinc-lead blast furnace

the top of the stack. Although this air reacts with the CO to form CO_2, the exothermic reaction keeps the temperature well above 950°C.

On passing from the top of the furnace, the zinc vapour is rapidly condensed or "shock cooled" by spraying with molten lead at about 600°C. The drop in temperature is so rapid that the CO_2 has no opportunity of causing the back reaction $Zn + CO_2 \rightarrow ZnO + CO$. At this temperature, zinc dissolves in the molten lead. This lead-zinc alloy is circulated through a separate cooler chamber at about 400°C, and the reduced solubility of Zn (about 2%) at this temperature causes the zinc to separate from the lead and float on top. The molten zinc is run off and the lead returned to the spray chamber. The zinc produced is about 98% pure with the main impurities being Pb (1–2%), Fe and Cd.

The purity of the zinc may be sufficient for use in hot-dip galvanising, but it is further refined to 99.99% for producing zinc-base die casting alloys and high-quality brass. This refining may be carried out by reflux distillation or by electrolysis of aqueous solution.

Pyrometallurgical Refining of Zinc

Blast furnace zinc may be refined by a process of fractional distillation. The refining process relies on differences in volatility—lead and iron being less volatile than zinc, and cadmium being more volatile. This process of fractional distillation is called *refluxing* and is a two-stage operation, two distillation columns being used (fig. 2.25). In the first stage, the Zn and Cd are

Fig. 2.25 Diagrammatic representation of zinc refining

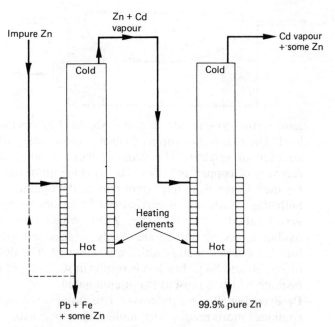

distilled from the Pb and Fe: then the Cd is evaporated from the Zn in the second stage by further refluxing and condensation.

In practice, the vapours evolved from the heated metal are passed up a column in which a temperature gradient is maintained, the temperature at the top being just sufficient to vaporize the Zn and Cd. As the vapours move up the column, the proportion of Pb and Fe continuously decreases since the temperature becomes progressively lower. The reverse happens to the condensate which flows down the column. The enriched Zn–Cd vapour is allowed to condense and is transferred to the second column, in which the temperature is lower than in the first, so that the Cd evaporates from the Zn and is subsequently recovered by condensation.

The columns contain refractory trays which help to give good contact between liquid and vapour. The zinc can be purified to 99.995%, the lead, iron and cadmium content together totalling less than 0.004%.

Pyrometallurgical Refining of Lead

The crude lead from the blast furnace (called base bullion) contains small amounts of many other elements. Silver is usually present in amounts which warrant its recovery, while varying amounts of Cu, Sn, As, Sb, Bi and Fe may also be present. Most of the impurities are held in suspension rather than in alloy form. Hence in the preliminary stages of refining a simple mechanical treatment suffices for their partial elimination. The various operation stages are described below.

Drossing The crude lead is held molten at about 350°C and agitated with a mechanical stirrer. This allows impurities which are less soluble at the lower

Fig. 2.26 Lead refining kettle

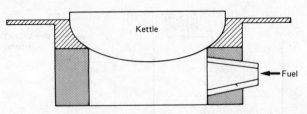

temperature to separate (e.g. As, Sb, Sn, Fe) together with a fair amount of lead. The removal of copper (which is immiscible with lead) is helped by the addition of sulphur. The dross is then skimmed off and treated for the recovery of copper and lead. The working up of the many drosses and slags for their values is an important part of the operations in a lead refinery.

Softening   Softening is carried out in an open-topped saucer-shaped steel vessel called a "kettle" (fig. 2.26), which presents a large surface for oxidation. As, Sb and Sn are slowly oxidised out with air and litharge PbO, forming a series of slags which are skimmed. The decreasing of the amounts of As, Sb and Sn to low levels results in softening of the lead. However, any bismuth which is present is not eliminated.

Desilverising (Parke's process)   This process is based on the fact that silver combines more readily with molten zinc than with molten lead, the Zn–Ag alloy being nearly insoluble in lead when sufficient zinc is present to saturate the lead. The softened lead bullion is heated to about 550°C and enough zinc added (2%) to saturate the lead and combine with the silver. Agitation and cooling to about 480°C results in a solid crust of Zn-Ag compound which is skimmed off. The silver and zinc are then separated by distillation after treatment in a press to remove excess lead withdrawn during skimming.

Dezincing   Lead ceases to be known as bullion when the silver has been removed, but further processing is necessary to remove the zinc (about 0.5%) remaining in the lead after desilverisation. This is achieved by one of the following methods:

1   Oxidation as in softening.
2   Conversion to zinc chloride by chlorine.
3   Vacuum distillation which recovers the zinc for re-use.

Up to this stage, bismuth is not eliminated, and electolytic refining was formerly carried out when the bismuth content was high. However, bismuth can be separated during the pyrometallurgical refining of softened and desilverised lead by the addition of a calcium-magnesium alloy to combine with the bismuth to form a solid compound which separates out. The excess Ca-Mg alloy can be removed by oxidation, while bismuth may be recovered from the skimmings.

Final purification of the lead is achieved by pumping the metal to another kettle in which air agitation is carried out. The dross which contains all the residual traces of Zn, As and Sb is skimmed off, leaving 99.99% pure lead which is cast into bars.

Cupellation   The Ag-Zn-Pb crust obtained from the desilverisation operation contains up to 25% Ag and its recovery is profitable. The zinc is first removed by distillation at 1400°C in graphite retorts. Recovery of silver from the retorted metal takes place by cupellation, the molten bullion being exposed to an air blast on the hearth of a small reverberatory furnace. The lead is oxidised to PbO which is run off, leaving the silver behind which remains unaffected by the air blast. The silver obtained assays about 99% and this can be raised to about 99.5% by a second cupellation with the addition of potassium nitrate.

2.10 Pyrometal-lurgical Extraction of Tin

Tin occurs naturally as the oxide, which can be fairly easily reduced by carbon. However, the reduction is somewhat complicated by the fact that tin oxide is amphoteric, and tends to enter acidic slags as a silicate and basic slags as a stannate.

Ores

The only ore of tin is cassiterite or tinstone SnO_2, which is chiefly found in alluvial deposits formed from vein, or lode deposits by weathering over a very long time. Although alluvial deposits are not contaminated with other minerals in the same way as vein deposits are, the tin content is still only about 0.02%. The high price of tin (approx. £9000 per tonne, 1982) stems from the leanness of the ore.

Most of the world output of tin is extracted from the alluvial deposits in S.E. Asia (Malaysia, Indonesia and Thailand).

Ore Preparation

Cassiterite is commonly recovered from alluvial deposits by dredging. The fine tin-bearing material is then subjected to jigging and tabling, exploiting the high relative density (7.3) of cassiterite. The tin content of the concentrate is raised to about 60%.

Extraction

The tin concentrate is smelted with powdered anthracite coal and limestone flux in a reverberatory furnace at about 1200°C—the high temperature being necessary to achieve a fluid slag. Reduction takes place readily, the molten tin flowing continuously into a ladle, from which the slag overflows into a granulating pot. A considerable amount of tin (up to about 35%) enters the slag, which is re-smelted with a larger proportion of reducing agent and lime. The slag resulting from this re-smelting may still contain enough tin to warrant further smelting. The tin extracted in this way usually requires little refining, the main impurity being a small amount of iron.

Pyrometallurgical Refining of Tin

The refining of tin may involve liquation and/or drossing (or "boiling").

Liquation is the separation of one metal from another with a different melting point. The impure tin is melted in a small reverberatory furnace with a sloping hearth. The tin, because of its low melting point (232°C), trickles down the hearth leaving behind an unmelted alloy of tin and impurities (e.g. Fe, Cu). This impure alloy is subsequently smelted to recover more tin.

Drossing (boiling) exploits the fairly low affinity of tin for oxygen. The tin is melted in a steel vessel and a current of air or steam is blown into the molten metal. Base metal impurities (e.g. Fe, Zn) are oxidised and rise to the surface to be skimmed off as a dross.

The purity of the tin produced is about 99.9%.

2.11 Pyrometallurgical Extraction of Nickel

In the extraction of nickel from sulphide ore, the first steps are similar to those in the pyrometallurgical extraction of copper from sulphide ore. In fact, because the occurrence of nickel sulphide is accompanied by copper sulphide, the separation of these two sulphides is the main problem in the extraction.

Ores

The most important ore of nickel is pentlandite, in which nickel exists as a mixed nickel-iron sulphide, along with copper sulphide, and small amounts of cobalt and the precious metals platinum, palladium and gold.

The most important source is the sulphide ore found in the Sudbury region of Ontario, Canada and which accounts for about 50% of the world output of nickel. (Nickel also occurs as a hydrated nickel magnesium silicate in Cuba and New Caledonia; this ore does not contain any copper or precious metals.)

Ore preparation

The sulphide ore, containing 1–2% Ni, is treated by flotation to give a concentrate of nickel sulphide, along with sulphides of iron and copper.

Extraction

Several extraction methods are in use for nickel, and one process, starting from sulphide ore, is outlined below.

The first step, consisting of partial roasting and smelting, resulting in a matte of nickel and copper sulphides with some iron sulphide, is similar to that involved in copper extraction. Unlike the production of copper, this matte cannot be completely converted to nickel metal because the reaction between nickel sulphide and nickel oxide has a positive free energy change at 1200°C. However, during conversion small amounts of metallic nickel and copper are produced and form an alloy, which collects most of the precious metals present. Also the FeS present is oxidised to FeO and slagged off by the addition of SiO_2. Thus, the converting operation results in a matte of nickel sulphide and copper sulphide, together with a small amount of Ni-Cu alloy. The converted material is then cooled slowly through the temperature range 900–400°C, which allows the Cu_2S to separate first, then the Ni-Cu alloy followed by the Ni_3S_2. The solidified material is crushed and ground and the sulphides of nickel and copper are collected separately by froth flotation. The magnetic Ni–Cu alloy is treated for recovery of the platinum group of metals, while the copper sulphide concentrate is used for copper extraction. The nickel sulphide concentrate is sintered to produce NiO, which is reduced to crude nickel and refined electrolytically or by the Mond process.

Electrolytic Process

The nickel oxide sinter is crushed, mixed with coke, and charged to a reverberatory furnace in which it is reduced to crude nickel, containing about 95% Ni, which is cast into anode slabs. The cathodes consist of thin sheets of pure nickel and the electrolyte is pure nickel sulphate solution acidified with sulphuric acid. During electrolysis, the precious metals do not dissolve, but collect as a sludge at the bottom of the cells. This sludge is periodically removed and treated for recovery of gold, silver and platinum metals. The purity of the electrodeposited nickel is about 99.9%.

Mond Process

In this process (fig. 2.27) the nickel oxide sinter is first roasted to ensure complete removal of any residual sulphur. The oxide is then reduced by water gas (a mixture of carbon monoxide and hydrogen) at 400°C in large cylindrical vessels, about 40 m high, 2 m in diameter, known as *reducers*:

$$NiO + H_2 \rightarrow Ni + H_2O$$

The impure nickel is reacted at 50°C with carbon monoxide to form a volatile carbonyl $Ni(CO)_4$. This reaction is conducted in tall towers called *volatilisers*, which resemble the reducers. The $Ni(CO)_4$ vapour is then passed to decomposing towers at 180°C, where the carbonyl decomposes and the nickel is

Fig. 2.27 Diagrammatic representation of Mond process

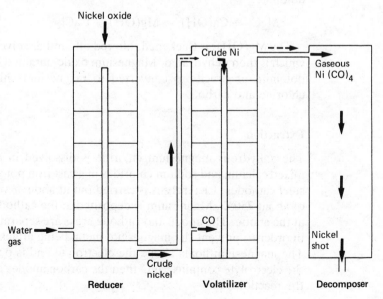

deposited on nickel shot circulating in the tower, and which provide centres on which the nickel can nucleate and grow:

$$Ni(CO)_4 \rightarrow Ni + 4CO$$

The liberated CO is passed back to the volatilisers to form more $Ni(CO)_4$. A constant volume is kept in the decomposers by providing an overflow pipe at the top, through which the final sized pellets pass. Impurities, e.g. Fe, Co and precious metals, present in the crude nickel entering the volatilising towers remain as a residue in the towers.

Monel metal, an alloy of 70% Ni and 30% Cu, is sometimes made directly. The matte, consisting of the sulphides of nickel and copper, is fully roasted to oxide by sintering and then smelted with carbon in a blast furnace, the two metals being reduced simultaneously to form the alloy.

2.12 Extraction of Magnesium

Electro-winning of magnesium is carried out using a fused chloride electrolyte at about 750°C.

Ores

Magnesium is extracted from deposits of magnesite ($MgCO_3$), dolomite ($MgCO_3.CaCO_3$), brucite ($MgO.H_2O$), carnallite ($MgCl_2.KCl.6H_2O$), and from sea water (containing about 0.13% Mg as $MgCl_2$).

Good-quality magnesite is found in Greece, Austria, U.S.A. and Canada. Dolomite sources are widespread.

Ore Preparation

Magnesium hydroxide is precipitated from sea water by addition of calcined dolomite:

$$MgCl_2 + Ca(OH)_2 \rightarrow Mg(OH)_2 + CaCl_2$$

The $Mg(OH)_2$ is thickened, filtered off, and dissolved in HCl to give $MgCl_2$ which is then dehydrated. Magnesium oxide obtained by calcining magnesite, dolomite or brucite is converted to magnesium chloride by heating with chlorine and carbon.

Extraction

The anhydrous magnesium chloride is dissolved in an electrolyte of fused mixed calcium and sodium chlorides in a cast iron pot with carbon anodes and steel cathodes. Electrolysis is carried out at about 7 volts with a temperature of about 750°C. Magnesium is deposited at the cathode and chlorine evolved at the anode. The anode and cathode areas are separated by a ceramic curtain in order to prevent the magnesium metal being attacked by the chlorine gas. The magnesium floats above the electrolyte and is periodically ladled off. If the electrolyte contains MgO then the carbon anodes are slowly consumed by the reaction.

$$C + MgO + Cl_2 \rightarrow MgCl_2 + CO$$

The energy consumption is about 20 kW hours per kg of magnesium produced.

2.13 Extraction of Titanium

Pyrometallurgical extraction of titanium presents considerable difficulties due to the great affinity of the molten metal for electro-negative elements such as carbon, nitrogen and oxygen as well as hydrogen.

Ores

The main ore is rutile TiO_2, which is nearly nonmagnetic. Titanium also occurs as ilmenite $FeO.TiO_2$, but this is used mainly for the production of TiO_2 pigment. Ilmenite has a high magnetic susceptibility.

The main source of rutile is Australia.

Ore Preparation

Finely divided rutile may be subjected to gravity separation to remove impurities such as SiO_2, as well as magnetic separation to remove any ilmenite that may be present.

Extraction

Large-scale production of titanium is achieved by the reduction of $TiCl_4$ by a reactive metal (e.g. Mg or Na).

In the *Kroll process*, purified rutile is ground with coke and briquetted using pitch as a binder. The material is then chlorinated at 800°C in a steel vessel to produce $TiCl_4$:

$$TiO_2 + 2Cl_2 + 2C \rightarrow TiCl_4 + 2CO$$

The $TiCl_4$ produced is a yellow liquid containing small amounts of impurities such as Fe, Si and Mg. These impurities are removed by carrying out fractional distillation in the presence of copper which acts as a reducing agent.

$TiCl_4$ vapour is then made to react with molten magnesium at 750°C in the presence of an inert gas (argon or helium) in a steel reaction vessel, thus

$$TiCl_4 + 2Mg \rightarrow Ti + 2MgCl_2$$

The $MgCl_2$ is leached away and excess magnesium removed by distillation.

The titanium is obtained as a spongy mass which must be consolidated into ingot form by melting. Because the molten metal is extremely reactive towards oxygen, nitrogen and hydrogen, melting must take place under vacuum. The consumable electrode arc type of melting furnace is used, the spongy metal being pressed into round or square electrode form. The electrode melts into a water-cooled copper crucible, but the titanium does not weld to the crucible because of the high thermal conductivity of copper. The molten titanium is then cast into an ingot under high vacuum.

The *I.C.I. process* is similar but uses sodium as the reducing agent instead of magnesium. The $TiCl_4$ is reduced by reaction with liquid sodium on solid sodium chloride. The reaction vessel, containing NaCl powder under an inert gas, is heated to 600°C to start the process and the temperature is then maintained by the heat released from the reaction

$$TiCl_4 + 4Na \rightarrow Ti + 4NaCl$$

Molten sodium is fed from the top and the inert gas stream carries the $TiCl_4$ upwards together with the NaCl powder. Some of the powder product containing Ti and NaCl is removed at intervals, and heated to 800°C out of contact with air to consolidate the titanium metal. After cooling, the NaCl is leached out with water.

2.14 Hydrometallurgical Extraction of Copper

In addition to the sulphide ores already dealt with (see section 2.8) copper also occurs in the form of low-grade oxidised ores, e.g.

Cuprite Cu_2O (reddish in colour)
Malachite $CuCO_3.Cu(OH)_2$ (greenish in colour)
Azurite $2CuCO_3.Cu(OH)_2$ (bluish in colour)

These ores are nearly always treated by hydrometallurgical methods.

The ore, in a fine particle size, is leached with dilute sulphuric acid, either in heaps of low-grade material or by agitation if of higher grade. The resulting leach liquor, after separation from insoluble solids, is treated as follows:

1 If the concentration of copper is fairly high the solution is given a preliminary purification treatment by ion exchange, then electrolysed using a pure copper cathode and a lead inert anode. The sulphuric acid formed during electrolysis is re-used as leachant. The copper produced is melted, cast and used for non-electrical purposes.

2 If the copper concentration is low, two processes are available. The older one is *cementation* of copper from solution on to scrap steel as represented by the reaction:

Fe	$+ Cu^{2+}$	$\rightarrow$	Cu	$+ Fe^{2+}$
(steel scrap)	(in solution)		(cemented metallic Cu)	(in solution)

The leach liquor is made to flow over steel sheets in large troughs, and copper is precipitated as a spongy deposit, which is removed from the steel by water sprays, collected and melted. The copper obtained is of low purity (85–90%) and requires purification. The chief contaminant is iron which is present up to about 2%.

The modern way of dealing with solution of low copper concentration is by solvent extraction. An organic solvent which is very suitable for extracting copper ions is hydroxy quinoline (C_9H_6NOH) dissolved in kerosene. The acidity of the leach liquor should be fairly low to allow ready removal of its copper content into the immiscible organic solvent. The copper is recovered from the solvent by reacting the latter with an aqueous solution strongly acidified with sulphuric acid.

The advantages of this method are

a) A solution of sufficient copper concentration for electrolysis is produced directly from dilute leach solutions.

b) The organic solvent has little solubility for impurity metals so that these remain in the leach liquor and the final electrolyte is of high purity. The copper is then recovered by electrolysis (see section 2.3).

2.15 Hydrometallurgical Extraction of Zinc

If the ore is smithsonite (or calamine) $ZnCO_3$, it is crushed and ground, then it may be subjected to a concentrating process. The next step is to leach the material in dilute sulphuric acid.

If the raw material is a sulphide ore containing ZnS and PbS (together with pyrites FeS_2), then differential flotation is used to obtain separate concentrates of ZnS and PbS. The ground ore in the flotation cell is first treated with sodium cyanide NaCN, to depress the action of the collector on the sulphides of zinc and iron, so that only the PbS is floated off. When this has been separated and removed, copper sulphate solution is added to the cell to activate the ZnS (by forming a very thin layer of copper sulphide on the surface) and the particles of ZnS can then be floated and separated from the iron sulphide.

The zinc sulphide is then roasted in fluo-solids roasters. The oxidation of ZnS to ZnO is almost complete, the sulphide-sulphur content being about 0.1–0.2%, with a sulphate-sulphur content of 1–2%. During the roasting, the iron sulphide present is oxidised to iron oxide, which then combines with zinc oxide to form zinc ferrite $ZnO.Fe_2O_3$.

The calcine is then reacted with hot dilute sulphuric acid solution in rubber or plastic-lined steel tanks. Zinc ferrite is readily soluble in the leachant. The solution is then neutralised with ZnO and purified. Formerly, iron was removed by precipitation as $Fe(OH)_3$, but the precipitate was bulky and difficult to wash free from occluded zinc sulphate. The modern method of iron removal is as basic iron sulphate which has a composition corresponding to the naturally-occurring insoluble basic iron sulphate known as jarosite. The basic iron sulphate precipitate is denser and easier to settle, filter and wash free from $ZnSO_4$ than $Fe(OH)_3$. The jarosite also collects co-precipitated arsenic and tin.

The solution is then clarified in thickeners and zinc dust added to precipitate metallic impurities including copper and cadmium. After filtering, the solution is ready for electrolysis. The need to achieve and maintain a very high standard of electrolyte purity dominates the electrolytic production of zinc. Zinc is electro-negative to hydrogen and electrodeposition of the metal is possible only if the overvoltage or resistance to liberation of hydrogen is high, which is so in the almost complete absence of impurities—especially As and Sb. It is vital, therefore, to use an extremely pure electrolyte to achieve satisfactory deposition of zinc.

Electrolysis is carried out in rectangular tanks made of pvc sheet supported by a steel frame. The cathodes are made of aluminium sheet and the inert anodes are of lead, to which about 1% of silver is added to reduce corrosion.

The electrolyte is water-cooled to remove the considerable amount of heat liberated.

The zinc tends to deposit as irregular outgrowths of metal at certain localised areas on the cathode. This phenomenon, known as "treeing", is minimised by the addition of glue to the electrolyte.

After stripping the zinc from the cathodes, the metal is washed, dried, melted and cast into slabs with a metal purity of 99.95 to 99.99%.

2.16 Electrolytic Refining

In electrolytic refining a soluble anode of the impure metal being refined is used, while the cathode is usually a thin "starting sheet" of the pure metal, although sometimes a blank of another metal is used. The electrolyte is a pure aqueous solution of a salt of the metal being refined; it must have high electrical conductivity and it is kept at a constant concentration.

The behaviour of the impurities in the anode is important. Metals *less noble* than the main metal go into solution in the electrolyte with it, but are not deposited on the cathode if the cell conditions are well chosen. Metals *more noble* than the main metal do not dissolve in the electrolyte, and remain behind on the anode or else fall to the bottom of the cell as anode slime, which may be collected in a fabric bag hung around the anode. These slimes may contain precious metals whose values warrant recovery.

In practice, the behaviour of impurities in the anode is not so simple, and depends to some extent on whether the impurities are in solid solution or combined with the main metal. Further, some of the main metal also enters the anode slime, because the uneven solution of the anode causes small pieces of the latter to fall off. In order to prevent co-deposition of other metals, the electrolyte will need regular purification because of the build-up of soluble impurities and a portion is drawn off for treatment, either continuously or periodically. Occasionally, impurities may form a small amount of precipitate in the electrolyte. In some cases the presence of other elements in the electrolyte may seriously affect the efficiency of the process and also the physical form of the deposit.

The use of a high current density should be avoided because it may cause co-deposition of impurities due to depletion in main metal ions of the electrolyte near the cathode, diffusion not being sufficiently rapid.

In electro-refining, the net chemical work is zero since the process only involves the transfer of pure metal from anode to cathode. The applied potential required to achieve this is low. The use of a warm electrolyte, which is adequately stirred, improves the efficiency of the process.

Electrolytic Refining of Copper

The electro-refining of copper is a very slow process (it may take up to about one month), but very pure copper is produced having high electrical conductivity. Such refining not only achieves complete recovery of any precious metals present, but also enables recovery as byproducts of impurities such as Se, Te, Bi, As, Sb, thus making the slow process economical.

The impure copper (which may have been fire-refined) is cast into anodes

Fig. 2.28 Walker
multiple system

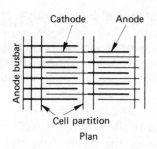

Plan

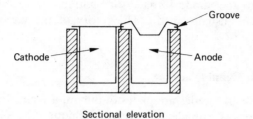

Sectional elevation

which are provided with lugs which serve both as a support in the tank and for making electrical contact with the current busbar. The cathodes consist of thin "starter sheets" of pure copper, about 1 mm thick, made by electroplating copper on to steel sheet and then stripping off the deposit. The cathodes are provided with hooks and are suspended from hanger bars in the tanks. In order to decrease the number of copper busbars for current carrying, the Walker system is used (fig. 2.28). In this arrangement the tanks are built in rows side by side, each pair of tanks having a side in common. The rods supporting the cathodes in the tanks rest on the partition making electrical connection by contact with grooves (known as Baltimore grooves) cut in the supporting lugs of the anodes in the adjacent tank. Thus, there is no need for intertank busbars, the current entering by a busbar at one of the row of tanks and leaving from the opposite end.

The electrolyte consists of pure copper sulphate solution acidified with sulphuric acid. The tanks are lined with lead or plastic material to withstand the corrosive effect of the acid electrolyte. Electrolysis is carried out at about 50°C in order to improve ion mobility and lower the cell resistance. On passing an electric current, copper is dissolved electrolytically at the anode and an equivalent amount of pure copper is deposited at the cathode. The reactions at the electrodes are as follows:

at *anode* $Cu - 2e \rightarrow Cu^{2+}$

(Under the conditions prevailing in this electrolytic cell the ions SO_4^{2-} and OH^- are not discharged.)

at *cathode* $Cu^{2+} + 2e \rightarrow Cu$

Impurities in the anode behave as follows:

a) Impurities which are less noble than Cu (e.g. Fe, Zn, Ni, Bi) dissolve in the electrolyte, but can be prevented from depositing on the cathode by keeping their concentration in the electrolyte low by subjecting the latter to purification treatment.

b) Impurities which are more noble than Cu (e.g. Ag, Au, Pt) do not dissolve and eventually fall into the anode slime, which is removed periodically and processed to recover the precious metals and also such elements as selenium and tellurium.

The anodes remain in the electrolytic cells for about a month when the remains (about 10% of the original weight) are removed, washed and returned to the anode furnace for melting and recasting. In this time two cathodes are usually made. The purity of the copper produced is about 99.98%.

Electrolytic Refining of Zinc

In this case the anodes are plates of the impure zinc requiring refining, while the cathodes are made of sheet aluminium. The electrolyte is pure zinc sulphate solution slightly acidified with sulphuric acid.

The details of the process are similar to the last stage of the hydrometallurgical extraction of zinc (see section 2.15).

Electrolytic Refining of Lead

The electro-refining of lead by the Betts process uses an electrolyte of lead fluosilicate $PbSiF_6$ ($PbF_2.SiF_4$), because lead forms insoluble salts with many acids including H_2SO_4 and HCl.

The lead is first fire-refined so that impurities total less than 2%; copper, in particular, should be low, otherwise it forms a hard insoluble skin on the anodes. The softened lead is cast into anodes, while the cathodes consist of thin sheets of pure lead. The lead fluosilicate electrolyte is acidified with hydrofluosilicic acid H_2SiF_6.

Impurities, including As, Sb, Bi and Cu as well as any precious metals present, remain in the anode slime, which is separated and treated to recover the various metals.

Tin is co-deposited with the lead because the two metals are very close in the electrochemical series. Tin may be removed from the lead by preferential oxidation either before or after electrolysis.

The purity of the refined lead is about 99.99%.

Exercises 2

1 Name the three main groups of extraction processes.

2 Explain what is meant by the following terms:
calcination, roasting, smelting.

3 *a*) Name four functions of a slag in molten metal processing.
b) Name four desirable properties of a slag.

4 Explain what is meant by the following terms as applied to slags:
a) basic, *b*) acidic, *c*) oxidising, *d*) reducing.

5 *a*) Explain the term "leaching" of a metal ore.
b) Name four types of leaching process

6 *a*) Name four processes of recovery of the valuable metal from leach liquor.
b) Describe one of the processes mentioned in *a*).

7 *a*) State Faraday's laws of electrolysis.
b) A current of 3.5 amps is passed for 100 minutes between platinum electrodes immersed in a solution of copper sulphate. Calculate the masses of products evolved at the electrodes. (Relative atomic mass of copper = 63.6 and of oxygen = 16; 1 Faraday = 96 487 coulombs).
[*Answer*: 6.92 g copper at cathode and 1.74 g oxygen at anode.]

8 Explain what is meant by the following terms:
a) enthalpy, *b*) entropy.

9 *a*) Explain what is meant by a "free energy diagram" (Ellingham diagram).
b) State the uses and limitations of the above diagrams in extraction metallurgy.

10 Complete the following table for the classification of iron ores:

Colour	Name of ore	Typical Fe content	Type of ore
Black			
Red			
Brown			
Yellowish brown			

11 *a*) Name four characteristics used to evaluate an iron ore.
b) Name four important sources of iron ore.

12 State the three important zones in the iron blast furnace, and name the refractories usually used to line the above zones.

13 With reference to the blast furnace smelting of iron ore explain what is meant by the following terms:
a) direct reduction, *b*) indirect reduction, *c*) the "coke rate".

14 *a*) Name two disadvantages of blast furnace smelting of iron ore.
b) Name two possible alternatives to the blast furnace production of iron.

15 Make labelled sketches to illustrate the stages of steelmaking by the BOS process.

16 Make a diagram to illustrate the removal of the common impurities during the blowing of oxygen in the BOS steelmaking process.

17 *a*) Make a simple sketch of an electric arc steelmaking furnace.
b) Name four stages into which electric arc steelmaking may be divided.
c) Distinguish between the oxidising slag and the reducing slag used in electric arc steelmaking.

18 Name three important sources of bauxite, and indicate the main impurities expected.

19 Give a brief account of each of the four steps in the Bayer process for the removal of impurities from bauxite.

20 Explain why the electrolysis of a fused salt is necessary for the production of aluminium.

21 With reference to the pyrometallurgical extraction of copper from sulphide ores:
   a) Give the names and formulae of two ores.
   b) Name two sources of copper sulphide ore.
   c) Give an account of each of the main stages in the extraction process.

22 Explain what is meant by a) blister copper, b) tough pitch copper.

23 Write an account of the blast furnace production of zinc and lead from a mixed sulphide concentrate.

24 a) Name the four main stages involved in the pyrometallurgical refining of lead.
   b) Write a brief account of each of the above stages.

25 Write an account of the production of 99.9% pure tin under the headings:
   ore, ore preparation, extraction, refining.

26 Write an account of the pyrometallurgical extraction of nickel from sulphide ore by the Mond process.

27 Write an account of the extraction of magnesium from sea water.

28 Write an account of the production of titanium by the Kroll process.

29 Name two methods for the recovery of copper from the leachant obtained after treatment of a low grade copper ore.

30 Write an account of the hydrometallurgical extraction of zinc from a sulphide ore.

31 Outline the differences in the cell conditions required for electrolytic extraction as compared with electrolytic refining.

32 Explain why the purity of the electrolyte used in the electrolytic extraction of zinc is so important.

Fig. 3.1 Cupola furnace

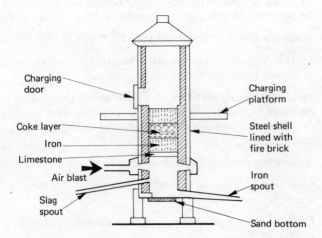

3 Melting and Casting of Metals

Straightforward melting operations (without any refining) are carried out in air, inert atmospheres or vacuum. Air melting is used for the less reactive metals and the low melting point materials, while the more reactive metals and high melting point materials are melted in inert atmospheres or vacuum. During air melting, considerable loss can occur by volatilisation of metals with low boiling points (e.g. Zn, Mg), while difficulties also arise when metals of widely differing melting points are being alloyed.

3.1 Melting Furnaces

Melting furnaces may be classified in a variety of ways. One classification uses the method of heating, distinguishing flame-fired and electric furnaces; another identifies shaft, hearth, rotary and crucible types. The furnaces vary widely in capacity and design. However, to achieve efficient melting it is essential to have effective transfer of heat from the heat source to the metal charge.

Shaft furnaces consist of a tall, cylindrical steel shell lined with refractory brickwork. The raw materials enter at the top of the furnace while the molten metal is tapped from the bottom.

The melting of pig iron in iron foundries is usually carried out in a furnace called a "*Ćupola*", which is a shaft-type furnace (fig. 3.1). The fuel used in the cupola is metallurgical coke and this, together with pig iron and limestone flux, is charged into the top of the furnace. The cold charge moves down the furnace shaft and meets an upward flow of hot gases, the close contact resulting in a high thermal efficiency. However, because the charge is in direct contact with both the fuel and the products of combustion, pick-up of carbon and sulphur from the coke takes place. Also the presence of coke results in a strongly reducing atmosphere. These factors restrict the cupola to use as a melting unit only—no effective refining can take place.

The vertical steel shell is lined with firebrick or rammed refractory material and mounted on a base-plate which is supported by four steel columns. The air-blast is admitted to the cupola through tuyeres arranged circumferentially near the bottom of the furnace, the tuyeres being connected to the air blower by means of a blast-box which encircles the furnace. Modern cupolas are usually of the "drop-bottom" type, i.e. they are fitted with hinged doors in the base-plate, and these can be lowered to allow solid residue from the furnace to be removed at the end of the melt. Before the cupola is charged, a bed of refractory material (e.g. sand) is built up on the drop bottom and made to slope towards the metal tap-hole. Also a slag-hole is provided below the tuyeres for the removal of slag at suitable intervals during the melt. A platform and charging hole are provided about 4 to 5 m above the tuyere level and here pig iron, coke, limestone and scrap iron are charged into the furnace.

Both cold- and hot-blast cupolas are in use. In the latter, the in-going air is pre-heated by the hot gases emerging from the cupola. Good-quality metallurgical coke is more difficult to obtain than hitherto so that the cupola has become less competitive regarding melting costs and electric melting is being increasingly used.

In **Hearth furnaces**, only the products of combustion come into contact with the charge—there is no direct contact between fuel and charge. Hence, contamination of the charge is decreased. The efficiency of heat transfer in these circumstances is enhanced by the hot gases resulting from combustion of the fuel being deflected (reverberated) by the arched roof of the furnace onto the charge. Heat transfer is therefore both by direct contact between the hot gases and the charge as well as by radiation from the hot arched roof and luminous hot gases. Furnaces based on this principle are called **reverberatory furnaces** (fig. 3.2) and may be fired by gas, oil or pulverised coal. The hearth of the furnace is rather shallow, but has a relatively large surface area, thus encouraging effective heat transfer.

Fig. 3.2 Reverberatory furnace

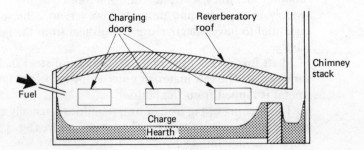

In the reverberatory furnace, good control can be exercised over the furnace atmosphere, and this represents an important advantage over the cupola. Because of the relatively high standard of cleanliness possible, this type of furnace is used for large-scale melting of metals and alloys of commerical purity. The pick-up of sulphur from combustion gases can be troublesome however, and in order to decrease contamination, melting is carried out under the protective covering of a molten flux or slag. Reverberatory furnaces vary widely in capacity up to a few hundred tonnes of liquid metal.

In **crucible furnaces** (fig. 3.3) the charge is isolated from both the fuel and the products of combustion. The crucible may be fitted with a lid, or the molten metal may be covered with a layer of flux, so that the risk of impurity pick-up is very much decreased. The metal to be melted is placed inside a crucible, the outer walls of which are heated using coke, oil or gas. The crucible may be of a refractory material (e.g. fireclay or graphite) or sometimes of metal (e.g. cast iron or steel). Crucible capacities range from a

Fig. 3.3 Crucible furnaces

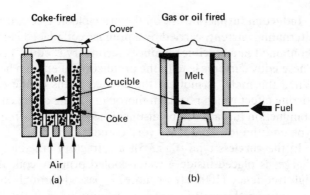

Fig. 3.4 A rotary melting furnace

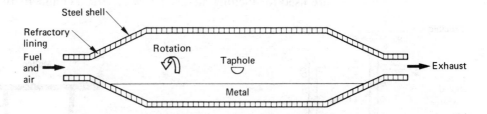

few kilogrammes to about 1 tonne. They are used mainly for melting small amounts of metal in a batch process. Consequently, the heat losses in the time interval between pouring one melt and charging the next are considerable so that thermal efficiency is low.

Rotary furnaces consist of a cylindrical steel shell, tapered at each end and lined with refractory material (fig. 3.4). The burners are positioned at one end, and at the other end there is an exhaust outlet, which is movable to allow charging of the furnace. The outlet is connected to a flue system. The furnace may be fired by gas, oil or pulverised coal and is mounted on rollers. When melting begins, the furnace may be rocked or completely rotated in the horizontal plane, the tap-hole in the side of the furnace being sealed. Rotation improves the transfer of heat because the metal is heated by conduction from the heated refractories as well as by radiation from the flame. The rotary motion also helps to achieve thorough mixing of the charge. Small gas-fired rotary furnaces are used for melting non-ferrous metals (e.g. brass) and have capacities of up to about 2 tonnes. Larger furnaces (capacity up to 20 tonnes) fired by pulverised coal are used for melting cast irons and may utilise a waste-heat recovery system in order to pre-heat the àir for combustion.

Electric furnaces are usually used where a high degree of control over melting conditions is required with minimum contamination and metal loss. Electric melting furnaces may be classified into induction, arc and resistance types. Arc furnaces are more suitable for the higher temperatures and where some refining is required, while induction furnaces are mainly used for straightforward melting and alloying. Electric resistance furnaces are sometimes used for melting small amounts of low melting point metals.

Induction furnaces work on the principle of electromagnetic induction. An alternating current, carried by a primary coil placed close to the metal charge (contained in a crucible), induces secondary ("eddy") currents in the charge. These eddy currents lead to the production of enough heat to melt the charge. When the metal is molten, the circular paths of the eddy currents cause a stirring effect which results in thorough mixing of the charge. Two main types of induction furnace can be distinguished—the high-frequency (or "coreless") type and the low-frequency (or "cored") type.

In the **coreless** type (fig. 3.5), a refractory crucible, containing the metal charge, is placed inside a water-cooled primary coil. An electric current of high frequency (1000 Hz or more) is passed through the coil, inducing eddy currents which result in sufficient heating to melt the charge. Such furnaces are used for melting metals of low electrical conductivity.

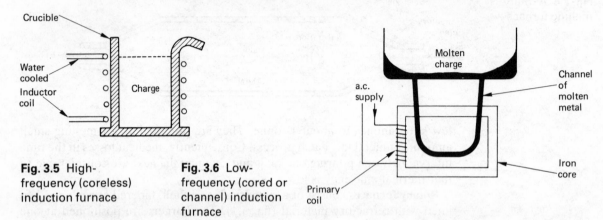

Fig. 3.5 High-frequency (coreless) induction furnace

Fig. 3.6 Low-frequency (cored or channel) induction furnace

In the **cored** (or channel) type of furnace (fig. 3.6), low-frequency (mains) a.c. is used and its applications include large-scale melting in iron foundries as well as for aluminium, copper and their alloys. The electrical circuit is similar to that of a transformer, in which two coils of wire are wound round a soft iron core. When an alternating current is passed through one coil, called the primary, an electromagnetic field is induced in the iron core, and this induces a current in the other coil, called the secondary. If the number of turns of wire in the secondary is greater than in the primary, then the potential difference across the terminals of the secondary will be greater than that applied to the primary. The primary winding of the furnace consists of a copper coil with a very large number of turns, while the secondary is a single loop. The loop is primed by filling it with molten metal, before the primary coil is energised. A very large secondary current is induced in the liquid metal, so that a flow of metal is set up around the loop. Solid metal is then added to the furnace crucible and melting of this charge proceeds.

This type of furnace is suitable for the continuous melting of alloys of similar composition, but is unsatisfactory for intermittent work with a variety of alloys.

Arc-furnaces may be subdivided into direct-arc and indirect-arc types. In the *direct-arc type* (fig. 3.7) an arc is struck from carbon electrodes onto the charge itself. The electrodes (usually three in number) are connected so that the current flows through the charge from one electrode to the next. Heating takes place both by direct conduction from the arc and by radiation from the furnace roof and walls. The furnace chamber is bowl-shaped and is fitted with a domed roof which carries the electrodes. The roof may be mounted on rails so that it can be moved aside while the furnace is being charged, or alternatively a charging door is fitted in that side of the furnace opposite to the spout.

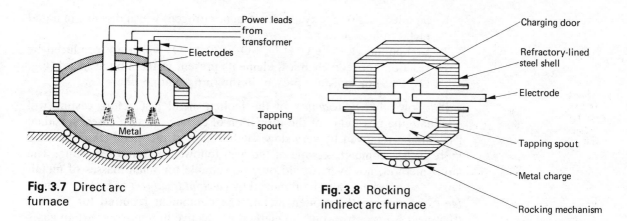

Fig. 3.7 Direct arc furnace

Fig. 3.8 Rocking indirect arc furnace

The *rocking indirect arc* furnace (fig. 3.8) consists of a cylindrical shell and has two electrodes mounted horizontally. The arc is struck between these electrodes so that the current does not pass through the charge, heat transfer being by radiation from the arc. The furnace is designed to gently rock during the melting and this helps the heat transfer and also mixes the charge.

3.2 Melting Practice

The main requirements of a molten metal ready for casting are:

1 A chemical composition complying with the specification.
2 A purity of the required standard.
3 A suitable casting temperature.

To achieve these requirements the metal should be melted as quickly as possible, using accurate temperature control and exercising proper precautions against avoidable contamination. The longer the melting time, the greater the opportunity for pick-up of impurities from plant, refractories and fuels, and the lower the throughput of metal, with resulting higher costs.

Since both oxidation and solution of gas in the metal increase with temperature, it is important to avoid overheating which might also cause the loss of volatile elements. The need for a large amount of super-heat can be avoided by pre-heating ladles and stirrers so that their chilling effect is reduced. Unnecessary turbulence of the melt will expose fresh metal surfaces which oxidise and form additional dross (oxide scum). Again the use of dirty, corroded scrap or small scrap with its relatively large surface area will increase the amount of dross produced.

It is often necessary to undertake the **removal of gases** from liquid metal since gas solubility decreases markedly on solidification. If an appreciable amount of gas has been dissolved by the metal during melting, this will cause porosity in the solidified metal because the gas will have difficulty in escaping completely. The more rapid the rate of solidification, the greater the risk of gas entrapment causing porosity in the solidified metal.

Gases in liquid metals may be of two types:

1 Simple gases (e.g. oxygen, hydrogen, nitrogen) which dissolve in liquid metal in atomic form.
2 Compound gases (e.g. H_2O, SO_2, CO_2, CO) which are produced by reaction between chemical elements present in the molten metal (e.g. O_2, C, S) or moisture present in the furnace atmosphere.

It follows that *deoxidation* of the molten metal just before casting will eliminate the possibility of the formation of certain reaction gases. Residual gas may be removed by very slow solidification of the metal (thus allowing more or less complete escape of the gas) followed by rapid remelting. The latter method, however, would only be suitable for small masses of metal. Dissolved gases may also be removed by *vacuum treatment* or by bubbling an *inert gas* through the molten metal. The equipment required for vacuum treatment is expensive, but the method is effective in removing certain gases (e.g. hydrogen) which have a rapid rate of diffusion. However, the method is not suitable for removing oxygen, which has a low diffusion rate. Bubbling an inert gas (e.g. argon) through a molten metal causes any dissolved impurity gas to diffuse into the bubble of inert gas. The bubbles rise to the surface and then escape, thus removing the impurity gas.

3.3 Alloying Methods

The charge required to produce a melt may contain virgin metal, secondary alloys and scrap in a wide range of possible combinations. The melting and alloying procedure depends on the percentage of various constituent metals, their relative melting and boiling points, physical form, solubilities and reactivities. When producing an alloy very rich in one constituent, the major metal (solvent) is first melted and then the minor constituent (solute) added. The solute metal may be added in a fine form because the solution rate is a function of the surface area. However, powder is liable to oxidation and compacting is usual before addition. When making an alloy of metals with widely different melting points, difficulties are encountered in obtaining complete solution without excessive metal losses due to volatilisation. Again considerable loss of low boiling point metals (e.g. zinc) can occur during melting in air.

The following are some of the methods used:

1   If the metals have similar melting points, they are mixed together and then melted, e.g. Pb and Sn; Cu and Ni.

2   If the solute metal has a low melting point compared with the solvent metal, it may be added before melting is complete because the alloying will lower the temperature required in the melt, and so help prevent excessive oxidation.

3   If the solute metal has a very high melting point, the solvent metal must be given an appropriate amount of super-heat. For example, when iron is added to copper, a temperature must be chosen that will give a good rate of solution while, at the same time, produce an acceptably low rate of oxidation.

4   Small percentages of high melting point metals may be added as "hardeners" which have lower melting points and dissolve more easily than the elements. A hardener contains a reasonably high percentage of the required solute, with the same solvent metal as that required in the final alloy. A composition for the hardener is chosen near to a eutectic if possible, in order to give a low melting point.

For example, P is added to Cu using 14% P/86% Cu hardener
Cu is added to Al using 50% Cu/50% Al hardener
Mn is added to Al using 25% Mn/75% Al hardener

5   Reactive metals may also be added as alloys but, if they have to be added in un-alloyed form, a common method is to use a rod with a perforated, expanded opening at the end. This is filled with the element and then thrust under the molten metal until the addition has melted and alloyed.

6   If the alloying metal is volatile, it is usually added as late as possible, e.g. in making brass the copper is melted first and the zinc added just before casting, making an allowance for loss by volatilisation.

3.4 Ingot Casting

A casting process involves pouring or "teeming" molten metal into a mould in which the metal solidifies. The shape into which the metal is cast varies according to the subsequent processing it will undergo. Two general types of cast products may be distinguished, namely ingots and castings.

Ingots are of relatively simple, symmetrical shape and are subsequently further processed by mechanical working.

Castings are finished articles and are often of complicated shape.

In ingot casting, metallurgical as well as economic considerations dictate that the molten metal should solidify quickly. To achieve rapid freezing, ingot moulds are made of metal and are either massive and of cast iron, or thin walled and of copper which is water-cooled. This ensures that the molten metal will chill rapidly on contact with the mould surface.

The molten metal may be transferred directly from the furnace or crucible into the mould, or it may be tapped from the furnace into a ladle and then teemed into the mould. The metal must be poured at a steady rate in order to avoid splashing and turbulence of the metal in the mould. Splashing results in nodules of frozen metal on the mould walls, while turbulence causes entrapment of air and other gases.

Fig. 3.9 Ingot casting methods

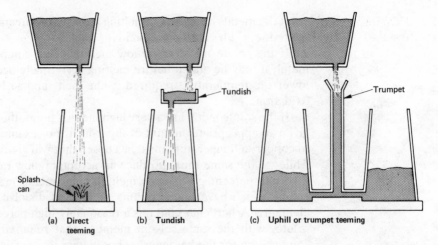

(a) Direct teeming (b) Tundish (c) Uphill or trumpet teeming

The molten metal may enter the mould at the top (top pouring) or from the bottom (bottom pouring) as shown in fig. 3.9.

Top-pouring (also called direct pouring or down-hill teeming) is the simplest and cheapest method, but gives poor control over the rate of teeming, so that splashing of the metal onto the mould walls may occur. These splashes freeze quickly and thus result in surface imperfections. Better control over teeming rate may be achieved by using a *tundish*, which is a refractory-lined vessel, placed above the ingot mould. A tundish also provides another opportunity for any slag or dross in the main stream of metal to rise to the surface, so preventing it from entering the mould. A better-quality ingot is produced by using a tundish, but more metal is lost, and also a higher pouring temperature is required in order to allow for the increased heat loss.

In **bottom-pouring** (also called uphill teeming or trumpet teeming) the molten metal is teemed into a central trumpet, from where it flows through runners to the bottom of a number of ingot moulds (up to eight). The metal rises slowly without turbulence into the mould. No splashing of molten metal occurs and a high-quality ingot with good surface finish is produced. However, bottom pouring is more costly because extra work is needed to set up the trumpet assembly, and the solid metal which remains in the runners results in a high loss of metal. A further disadvantage is the possible pick-up of refractory material from the runners and care is needed to prevent this.

Tenacious oxide films are formed by some alloys, especially those containing substantial amounts of aluminium (e.g. aluminium bronze), and in these cases it is most important to ensure quiescent conditions during the pouring. Under turbulent conditions, which may exist during conventional pouring, these oxide films would be entrapped in the cast and would badly affect the mechanical properties. Therefore, special casting techniques are used to avoid turbulence, e.g. the *Durville* casting process (fig. 3.10). In this method the ingot mould is attached to the crucible and slowly inverted so that the molten metal flows quietly into the mould, leaving the oxide films on the surface.

Fig. 3.10 Durville
casting method

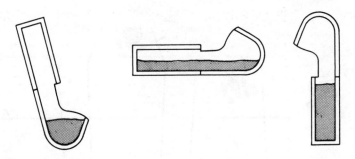

Fig. 3.11 Teeming
ladle

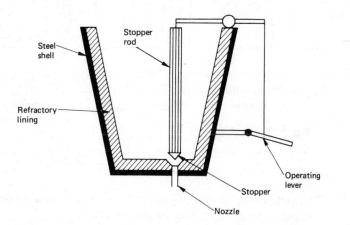

The **production of steel ingots** commences when the charge in the steel-making furnace is ready to be tapped into a ladle (fig. 3.11) which is a refractory-lined steel vessel capable of holding an entire furnace charge (up to about 300 tonnes). In order to prevent the molten slag on top of the steel from entering the ingot moulds, the steel is teemed from the bottom of the ladle. The outer shell of the ladle has an opening in the bottom, in which is fitted a refractory nozzle with a stopper. The stopper is carried on the end of a steel rod, protected by refractory sleeving which extends vertically through the molten bath of steel. The stopper rod is connected to a lever mechanism by means of which an operative controls the raising and lowering of the stopper and consequently the rate of teeming. As stated above it is of the greatest importance that teeming be carried out at the optimum rate: if too fast it will cause splashing in the mould, while if it is too slow it will result in premature freezing. The net result of incorrect teeming is an ingot of inferior quality, especially regarding surface finish.

At the end of the refining stage of steelmaking, the steel contains an excess of oxygen which exists as dissolved FeO. Steel in this condition cannot be satisfactorily cast because the FeO will react with the carbon present to form CO gas which bubbles out of the steel, and, if uncontrolled, would cause unacceptable porosity:

$$FeO + C \rightarrow Fe + CO$$

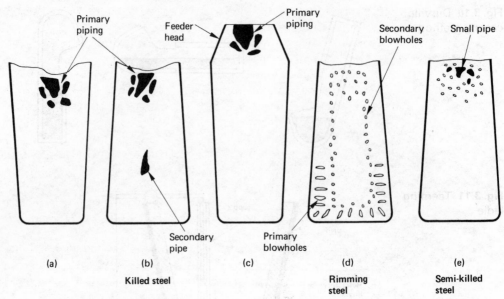

Primary piping — Feeder head — Primary piping — Secondary blowholes — Small pipe

Secondary pipe — Primary blowholes

(a) (b) (c) (d) (e)

Killed steel

Rimming steel

Semi-killed steel

Fig. 3.12 Steel ingot types

Depending on the degree of control exercised by deoxidation, three main types of steel ingot may be produced:

a) fully killed *b*) rimming *c*) semi-killed or balanced

In **fully killed steel** the oxygen is completely removed by adding deoxidisers, e.g. aluminium, ferro-silicon and ferro-manganese. The steel is said to be "killed" because it is quiescent during casting.

$$3FeO + 2Al \rightarrow 3Fe + Al_2O_3$$
$$2FeO + Si \rightarrow 2Fe + SiO_2$$
$$FeO + Mn \rightarrow Fe + MnO$$

The oxides formed by the deoxidation are insoluble in the steel and float to the surface to join the slag.

When the steel solidifies, a shrinkage cavity known as a "pipe" is formed in the top central part of the ingot. Because the surface of the pipe is oxidised it will not weld up during hot working and a considerable portion of the ingot would have to be scrapped. With the mould in the "big-end up" position (fig. 3.12*a*), the only piping is in the top central part of the ingot. If, however, a "big-end down" mould is used there may also be a secondary pipe (fig. 3.12*b*). Therefore big-end up moulds are usually used in the casting of killed steel. Piping is minimised by keeping the steel at the top of the mould molten for a longer time by fitting a *feeder head*, which is a steel shell lined with firebrick, on top of the ingot mould. The addition of an "exothermic powder" to the molten steel at the top of the mould generates further heat and so helps to maintain the necessary volume of molten metal. The pipe is then largely kept within the feeder head and a yield of about 80% good steel is produced from the ingot (fig. 3.12c).

In the case of **rimming steel** some deoxidation is carried out, but some oxygen remains in the steel in the ladle. After the steel has been teemed into big-end down ingot moulds and after some solidification has taken place, carbon monoxide is formed by the reaction $FeO + C \rightarrow Fe + CO$ and the gas bubbles out through the molten steel. Eventually the top of the ingot freezes over and the gas is no longer able to escape, so that blowholes are formed in the steel. These blowholes counteract the shrinkage which occurs during solidification and little or no pipe is formed (fig. 3.12d). The inside surfaces of these blowholes are not oxidised and, therefore, they will weld up during hot rolling. If the correct amount of deoxidiser has been added, and provided the temperature of the steel at the time of teeming is correct, the formation of the blowholes will not start until a solid *rim of pure iron* about 75 cm or more in thickness has been formed.

Rimming ingots give a high yield of good steel (about 90%) without any of the expense of using feeder heads.

In the case of **semi-killed or balanced steel** (fig. 3.12e) more deoxidation is carried out than with rimming steel, but not enough to completely "kill" the steel. If the carbon content of the steel is more than about 0.15%, then it is very difficult to achieve a good rimming action. Furthermore, steel should contain at least 0.25% carbon to justify complete deoxidation and the use of feeder heads to obtain an economic yield. Therefore steels with carbon contents between 0.15 and 0.25% may be cast into semi-killed or balanced ingots. Less gas is given off in the mould than with rimming steel, but enough bubbles form at the top of the ingot to counteract most of the shrinkage, so that usually only a very small pipe is formed. There is very little rise or fall in the top surface of the steel in the mould, hence the name "balanced steel". The yield of good steel is about 85%.

The exact shape and weight of an ingot depend on the subsequent processing it will undergo. For steel *casting*, the ingot moulds are made of haematite cast iron, and in cross-section they may be square for rolling into slabs, rectangular for production of plate and round for forging and tube making. The moulds have rounded corners to help avoid planes of weakness in the ingot structure. Alloy steel ingots for *forging* are commonly cast with fluted surfaces to acclerate solidification. Rapid heat extraction is also helped by making the mould walls very thick—up to about 25 cm thick. To facilitate the separation of ingot and mould, the latter is designed with a taper of about 6%. Ingots cast in big-end up moulds are lifted out from the moulds, whereas big-end down moulds are withdrawn from the ingots cast in them.

The principles of casting **non-ferrous ingots** are similar to those for steel ingots, except that the former are much smaller and the casting temperatures are much lower, so that most of the difficulties associated with size and slow cooling do not arise. Non-ferrous metals are cast into moulds which produce ingots nearer to the final shape required than in the case of steel. Thus ingots for rolling into sheet or strip are cast as slabs (about $1 \text{ m} \times 0.5 \text{ m}$); those for wire production as bars (called wire-bars, about 2.5 m long $\times$ 5 cm square), while for re-melting notched ingots are used (fig. 3.13).

Fig. 3.13 Non-ferrous ingot moulds

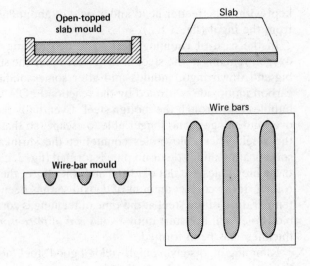

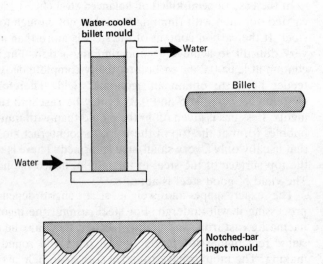

The types of moulds used include the following:

1 Horizontal cast iron moulds which are simple and cheap to use. However, the ratio of surface area to volume is high so that there is risk of oxide dross being trapped inside the cast product.
2 Vertical moulds which give a lower surface area to volume ratio, but at the expense of increased turbulence.
3 Water-cooled moulds which give a steadier cooling rate, thereby resulting in a better surface and fewer problems with dross and cracking.
4 Tilting moulds which may be used in similar fashion to the Durville process (see section 3.4, p. 78).

3.5 Ingot Defects

Ingot defects may be broadly classified into

a) Internal flaws caused by shrinkage.

b) Surface flaws which include cracks, porosity, ripples.

Shrinkage defects include primary piping, secondary piping and centre-line shrinkage [see section 1.5 *Structure*]. During the final stages of freezing there may be insufficient molten metal to feed into the spaces between grains resulting in porosity between the grains. Centre-line shrinkage may constitute a serious line of weakness in the ingot.

Surface defects may be classified as follows:

1 Defects due to an unsatisfactory rate of teeming and these may be subdivided as below:

(i) Splashing of molten metal causing drops to freeze on to the mould wall in advance of the rising column of metal.

(ii) Rippled surface due to teeming at too low a speed or too low a temperature.

(iii) Laps or folds resulting from the surface of the rising metal freezing and forming a layer. Laps are associated with bottom pouring too slowly or at too low a temperature.

2 Defects due to the use of worn moulds including the following:

(i) Fins or thin strips of metal protruding roughly at right angles to the ingot surface due to the molten steel having run into open cracks in the mould.

(ii) Surface crazing corresponding to crazing of the mould surface.

(iii) Scabs or bulges on ingot surface due to depressions in the mould wall.

3 Cracks arising from a number of causes including:

(i) Restriction of ingot skin on cooling (these cracks are called "hot tears").

(ii) Teeming at too high a temperature or too rapidly.

(iii) Mould temperature too high.

4 Surface pinholes caused by gas formed at the mould surface, e.g. gas liberated from a mould dressing or by moisture on the mould wall.

5 Dirty metal due to entrapment of slag or pieces of refractory.

3.6 Continuous Casting

Large-scale **continuous casting** was first used for aluminium and copper and their alloys so that most ingots in these materials are now continuously cast. Because of the much higher temperatures involved, the continuous casting of steel posed greater difficulties, and was a later development. However, at the present time (1982) about 50% of the world production of steel is continuously cast. The method is also applied to the alloys of zinc, magnesium, lead and tin.

There are two main continuous casting techniques, namely the vertical and horizontal methods. Non-ferrous metals are usually cast by the latter method, while the vertical method is used for steel. However, with improved technique, the horizontal method is being increasingly used for steel.

The continuous casting process involves pouring molten metal into a precisely constructed, oscillating, water-cooled mould, the internal dimensions of which govern the type of product being cast. The end of the solidified metal is gripped and pulled away from the mould, then passed through guide rolls until finally the product is cut to length.

In the **vertical casting method** (fig. 3.14) the molten metal is tapped from the furnace into a ladle (which may be provided with supplementary heating) and then passed into a refractory-lined vessel called a tundish, which may have several outlets. The metal then flows into the water-cooled copper mould so that heat may be rapidly conducted away. The mould is open-ended and is plugged at the beginning of each cast by a dummy bar on which the metal solidifies and which is drawn from the bottom of the mould. As the metal is poured, a cup is formed which is solid at the base and sides, but still molten in the centre. As the cup is drawn downwards its sides are being thickened by the freezing of the metal against the mould walls. The withdrawal speed is adjusted to keep pace with the thickening of solid metal. While this is going on, a constant level is maintained in the tundish. To prevent the metal sticking to the mould walls, a gentle oscillating movement is given to the mould, so giving a stripping action at the mould/metal interface. This is also helped by the use of a vegetable oil as a lubricant between the mould and the solid metal shell.

The core of the metal product remains molten for a considerable distance below the mould. To extract this remaining heat, water is sprayed onto the product between the mould and the withdrawal rolls. The product passes through pinch rolls in order to control the final dimensions and it is then bent into the horizontal plane. Solidification may not be entirely complete until the product is in the horizontal plane. The rapid cooling involved gives a pronounced pattern of solidification. The solid shell of metal first formed, consisting of very small equiaxed grains, is followed by a zone of columnar grains similar to the pattern which exists when a metal is cast from a high temperature into a cold metal mould. This unsatisfactory structure may also be accompanied by marked segregation of alloying elements (especially in copper-tin alloys.) However, immediately after withdrawal from the mould, the product may pass through electromagnetic stirring coils in order to decrease the tendency to form columnar grains and to encourage a finer grain size with less directionality.

The continuous casting of non-ferrous strip may be achieved by using a machine such as the "Hazelett" caster (fig. 3.15), which consists of two inclined, parallel, moving steel belts. Molten metal is poured into the highest point of the belt with solid strip emerging at the bottom end. The moving belts act as a frictionless mould.

Horizontal continuous casting machines (fig. 3.16) do not require the high structural steel support needed for vertical plant and therefore the capital outlay is much smaller. (Modern vertical continuous casting plants may use a curved mould which allows the required height to be much reduced). Also, controlled cooling of metal is easier and no bending of the product into the horizontal plane is involved. In the case of non-ferrous metals, the complete process of melting, alloying, holding and casting may be carried out with one unit, in which the molten metal may be protected by an inert atmosphere. This arrangement avoids the need for transfer of molten metal from one unit to another and thus reduces metal loss.

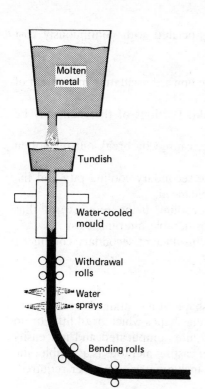

Fig. 3.14 Schematic diagram of a vertical continuous casting machine

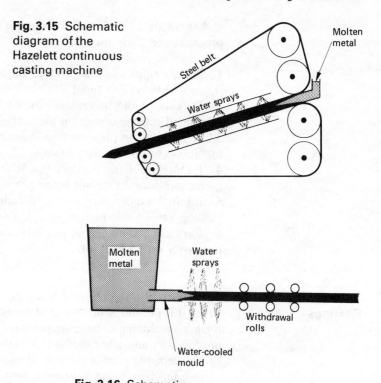

Fig. 3.15 Schematic diagram of the Hazelett continuous casting machine

Fig. 3.16 Schematic diagram of a horizontal continuous casting machine

The following **advantages** of continuous casting can be identified:

a) The elimination of the "ingot stage" with its re-heating and primary rolling leads to large capital savings even though the casting plant itself is more expensive than conventional casting equipment. Production costs are lower.

b) Less man-hours per tonne of product and a smaller working area are required.

c) Provided the process is under very close control, less discarded metal is produced than with conventional ingots, i.e. a higher yield is achieved.

d) Special shapes of small cross-section are readily produced.

e) Materials which are difficult to work may be readily cast to semi-finished products.

The **disadvantages** of continuous casting are:

a) The structure of the product may be coarse grained and highly directional, so that it may require considerable subsequent hot work to achieve satisfactory mechanical properties.

b) The control of the process parameters must be precise.

c) Any interruption of the process due to plant breakdown can lead to a serious "back-log" of molten metal.

A number of **defects** are commonly associated with continuously cast products and these include:

1 Surface ripples on the product due to the upward oscillating movement of the mould being too rapid.
2 Sticking of the metal shell to the mould. Portions of the shell may be entrapped causing a scab on the surface.
3 A lap may be produced on the surface, caused by break-out of molten metal due to poor casting practice.
4 Distortion of the product caused by the secondary cooling rate and the metal withdrawal rate not being properly balanced.
5 Internal cracks due to the secondary cooling being too severe, thus causing uneven contraction resulting in considerable internal stress.
6 Centreline shinkage or porosity due to insufficient secondary cooling or too fast a withdrawal rate.

3.7 Production of Castings

The ability of a molten metal to take the shape of a container (mould) into which it is poured is made use of in producing shapes which need little or no further machining. These shapes may be quite complicated and not easily produced by any other method. A variety of casting methods is available, the choice depending on a number of factors, including size, number required, surface finish and cost per component.

There was formerly some prejudice against castings for components subject to high loads or dynamic stresses. However technical progress made in the foundry industry has resulted in the availability of castings of much higher quality. Castings can now be produced to very high dimensional tolerances in complex shapes with weights ranging from a few grammes up to many tonnes, and in alloys impossible to fabricate by any other means.

Most metals contract (about 3 to 5% in volume) when cooling from the liquid to the solid state at the freezing point, and this must be taken into consideration when designing the casting. Any defects in a casting will probably be retained in the finished product, so that controlled freezing is essential.

The production of a casting may involve various moulding materials and methods.

Sand casting may use green (undried) sand or dried sand moulds. Green sand moulds are made of a mixture of silica sand bonded with clay to which is added a small amount of coal dust and cereal. It is most important that the quality of the moulding sand be maintained by regular checking of samples.

The making of a simple sand mould is illustrated in fig. 3.17. A replica (called a *pattern*) is first made from wood or mild steel; the pattern must be slightly larger than the required dimensions of the finished casting in order to allow for contraction of the casting during cooling. A mould is made by compacting sand around the pattern, and in order to facilitate removal of the pattern the mould is made in two (or more) parts. The pattern is laid on a moulding board and the bottom half (called the "drag") of a moulding box placed around the pattern. Moulding sand is introduced into the moulding

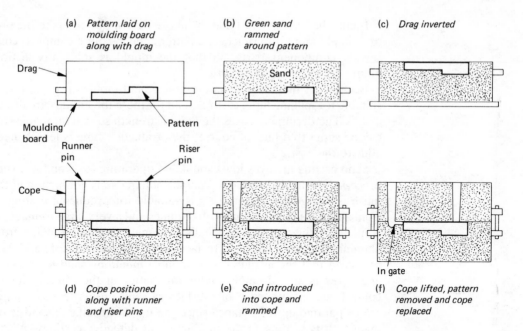

(a) *Pattern laid on moulding board along with drag*

(b) *Green sand rammed around pattern*

(c) *Drag inverted*

Drag

Moulding board

Pattern

Runner pin

Riser pin

Cope

Sand

In gate

(d) *Cope positioned along with runner and riser pins*

(e) *Sand introduced into cope and rammed*

(f) *Cope lifted, pattern removed and cope replaced*

Fig. 3.17 Production of a simple sand mould

box and the sand is pressed into contact with the pattern, either by hand or by machine, to produce the required compaction and strength. The excess sand is removed so that the sand is level with the sides of the moulding box. The drag is inverted and parting sand (dry, clay-free material) applied to the exposed surface so that the upper half (the "cope") of the mould will not adhere to it when this is made. The top half of the moulding box is now placed in position, the two halves being located by tapered pins and sockets. Runner and riser pins are put into the cope and a small quantity of moulding sand pressed around them. The runner receives the molten metal while the riser acts as a reservoir from which molten metal can feed back into the casting as it solidifies and shrinks. The riser is located at the highest point of the casting and its volume should be such that it will be the last portion of the casting to solidify. Moulding sand is now introduced into the cope and pressed around the pattern, runner and riser pins.

If hollow sections are required in the finished casting, then cores are placed in the mould cavity. Cores are separately moulded and must be carefully placed in position so that the molten metal will flow around them. The ramming of the sand around the pattern must not compact the mould too much, otherwise escape of gas is prevented. The mould may also be vented with fine holes to allow gas to escape.

The pattern is now removed by gently lifting the cope, which is then carefully replaced above the drag. The finished mould is now ready to receive its charge of molten metal which is poured in through the runner until it fills the riser. After the casting has solidified and cooled sufficiently, it is knocked out of the mould and most of the sand returned for further use. The casting is then "fettled", i.e. the runners and risers are cut off and the surface cleaned by wire-brushing, grinding or shot blasting.

It must be remembered that the above description refers to the preparation of a simple mould. In practice, a pattern may be quite complex, consisting of several component parts, so that the mould is split into sections and a multi-part moulding box is used.

Because of their fragile nature green sand moulds (i.e. undried) are not suitable for making large castings, and for these the moulds are dried at about 250°C. The drying increases the mould strength so that large masses of metal can be supported but, of course, the production time is lengthened and this adds to the cost.

Sand casting involves low capital and operating costs and is a very flexible process, but the dimensional accuracy obtainable is not high, and the surface finish is not good. The process is suitable for producing castings weighing from about 100 grammes up to 100 tonnes and a very large tonnage of castings is made by the method, in metals with melting points ranging from that of aluminium to that of steel. In fact such vast tonnages of sand castings are produced that the availability of suitable **moulding sand** for making moulds and cores is an extremely important matter in the foundry industry. The control and maintenance of satisfactory sand properties and characteristics are of paramount importance since the use of unsuitable sand is one of the main factors leading to the production of defective castings. A consistently acceptable standard of quality can only be achieved by good sand control. Most of the sand can be re-used after casting, but care is necessary to segregate sands to which special additives have been made.

Sand occurs naturally as deposits of two different types:

1 Sand with high silica content and free from clay. This material is used as a base for preparing synthetic sands by making various additions, depending on the intended use. Green sand is made by adding clay (5–10%) and moisture (3–4%) in amounts depending on the application. Synthetic sands can be prepared to a uniform composition to give reproducible results.
2 Naturally bonded sand containing a proportion of clay, known as natural green sand. The properties of this material are rather variable depending on the content of clay (usually up to about 10%), water (3 to 8%) and organic matter.

Coal dust (4–5%) used to be added to moulding sands to improve the surface finish of the casting by preventing the adhering of sand particles but with the advent of new additives the practice is no longer so widespread. The making of easily removable cores was formerly achieved by using linseed-oil bonded sand which needed a long curing time, but quicker curing resin-bonded sands are now used.

The **properties of a moulding sand** depend on the chemical nature and physical characteristics of the minerals in the bonding material. Hence the size of the silica grains, the clay, moisture and additive contents have a marked effect on the properties. The important characteristics of a moulding sand include the following:

1 *Moisture content*, which should be in the range 4–7%. A higher content promotes metal-mould reactions and also lowers the permeability of the sand, resulting in castings with numerous surface defects. High moisture content also causes swelling of the sand, thus affecting the dimensional accuracy of the casting. Moisture control is of paramount importance in ensuring good-quality moulds.

2 *Bond strength*, which determines the behaviour of the sand mould under stress and also the ease of removal of patterns from the mould. The mould must be strong enough to support the metal mass and also accommodate the changes due to temperature alteration.

3 *Plasticity*, which is a measure of the mouldability of the sand, i.e. its ability to change shape under load and to retain that shape after load removal. Low plasticity leads to unsatisfactory ramming of the sand and also hinders removal of the pattern after moulding.

4 *Permeability*, which is a measure of the ease with which gases may pass through the rammed material. A sand with a high permeability is called an "open sand", and one with low permeability is known as a "close sand". The venting qualities of moulds and cores depend on this property.

5 *Refractoriness*, which is the property enabling the sand to resist the adverse effects of heat and prevent the sand from fusing at the mould surface. Any "burning on" of the sand to the casting prevents easy removal of the sand and results in a poor surface on the casting.

Satisfactory **quality control** of moulding sand involves the regular testing of samples of the sand. The representative sample should be taken at the moulding station, avoiding sand which has been air-dried or spilled. A convenient mass of sand to enable all the various tests to be carried out is about 2 kg, and this should be collected in a sealable container. To avoid loss of moisture there should be no delay between weighing the sand for the various tests and the actual testing. For each of the various properties being determined the average of three tests, which are in reasonable agreement, should be taken.

For rapid routine control of **moisture content** use is made of the "Speedy Moisture Tester" (supplied by Thomas Ashworth and Co, Burnley, Lancs) which is illustrated in fig. 3.18. A fixed mass of sand is placed in the body of the container along with one measure of calcium carbide. The container is then shaken to allow the moisture present in the sand to react with the calcium carbide to form acetylene gas proportional in volume to the moisture content:

$$CaC_2 + 2H_2O \rightarrow Ca(OH)_2 + C_2H_2 \uparrow \text{ (acetylene)}$$

The gas pressure is indicated on a pressure gauge which is calibrated directly in percentage moisture.

Other properties of the sand are assessed by carrying out various tests on standard cylindrical test pieces rammed to a specific degree, with diameter 5.08 cm and length 5.08 cm.

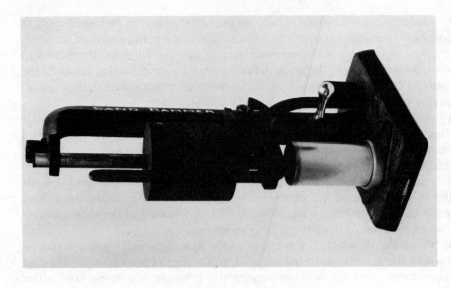

Fig. 3.19 Ridsdale-Dietert sand rammer (*courtesy* Ridsdale & Co Ltd)

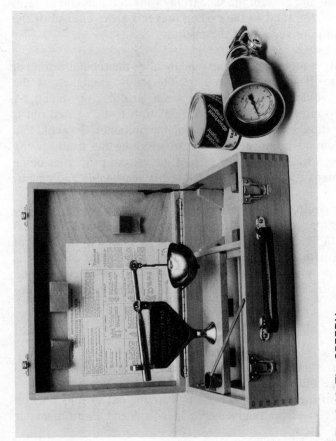

Fig. 3.18 The SPEEDY moisture tester (*courtesy* Thomas Ashworth & Co Ltd)

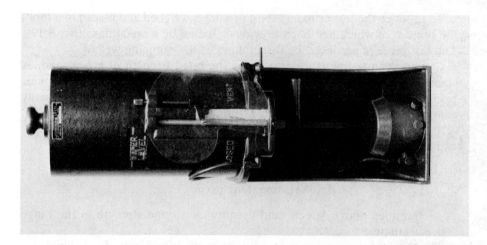

Fig. 3.22 Ridsdale-Dietert permeability apparatus

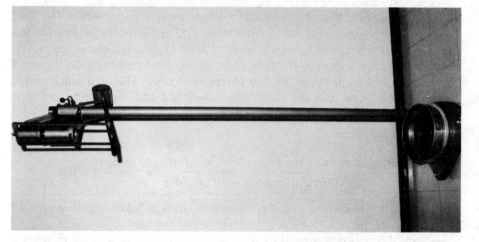

Fig. 3.21 Shatter test apparatus (*courtesy* Ridsdale & Co Ltd)

Fig. 3.20 Sand compression tester (*courtesy* Ridsdale & Co Ltd)

To produce the test cylinder, 160 g of sand is weighed and placed in a tube, the bottom of which had been previously sealed by a metal base (fig. 3.19). The test piece is produced by three blows of the ramming weight.

Compression testing is used to assess the bond strength of green sand. A standard cylinder of sand is prepared as described above—the test-piece being stripped from the tube on a stripping post. The test-piece is gradually compressed between the plates of a spring balance compression tester (fig. 3.20). The maximum load in gramme-force is noted at the point of collapse and the bond strength is calculated from the formula:

$$\text{Bond strength} = \frac{\text{Maximum load (gf)}}{\text{Cross-sectional area (cm}^2)}$$

As determined above, green sand usually has a bond strength in the range 500–850 gramme-force/cm^2.

The test can also be carried out on other sands including dried sand.

The **plasticity** of a sand is assessed by measuring the **shatter index**. The sand test-piece is positioned at the top of a tower 1.83 m high and ejected from the specimen tube so that it falls on to an anvil head 75 mm in diameter (fig. 3.21). On impact, the test-piece shatters, some of the sand remaining on the anvil and the rest being projected on to a 14 mm mesh sieve. The sand which passes through the sieve into the sieve pan is weighed and the shatter index is calculated as follows:

% shatter index =

$$\left(\frac{\text{Original mass of sand cylinder} - \text{Mass of sand in pan}}{\text{Original mass of sand cylinder}}\right) \times 100$$

The shatter index and green compressive strength taken together give an indication of the deformation characteristics of the sand.

Permeability is determined by measuring the rate of flow of air through a standard rammed test-piece prepared in the usual way.

The sand test-piece, in its container, is placed on the sealing device of a Ridsdale-Dietert permeability apparatus (fig. 3.22). The bell is raised and then allowed to fall until 2000 cm^3 of air has passed through the sand. The pressure is noted on the manometer and the time taken for the passage of the air measured.

Permeability is then calculated from the formula:

$$P = \frac{Vh}{p \times a \times t}$$

where P = permeability number
V = volume of air passing through test-piece in cm^3
h = height of test piece in cm
p = pressure of air in cm water gauge
a = cross-sectional area of test piece in cm^2
t = time in minutes.

Using the experimental conditions described above,

$$V = 2000\,\text{cm}^3 \qquad h = 5.08\,\text{cm} \qquad a = 20.268\,\text{cm}$$

The test may be simplified by standardising the manometer pressure at 10 cm and taking the time (in secs) for 1000 cm³ air to pass through the sand.

Other types of permeability testing apparatus are available including direct-reading instruments.

In the **carbon dioxide moulding process**, fine silica sand is mixed with a small amount of sodium silicate and the mixture is placed in the moulding box and rammed in the same way as green sand except that more vent holes are necessary. After moulding, CO_2 gas is blown through the sand for several seconds, and this results in a chemical reaction between the CO_2 and sodium silicate with the formation of sodium carbonate and silica gel, which bonds the sand grains together. The main disadvantage of this process is that the sand cannot be readily re-used. Moulds made by this method are strong and abrasion-resistant so that the process is suited to mechanised production. An important development is the use of cold setting resins instead of silicate and this eliminates the need for gassing with CO_2. In the case of some self-setting processes, the use of moulding boxes can be dispensed with. The use of these resins has also made core making much easier.

Fig. 3.23 Shell moulding

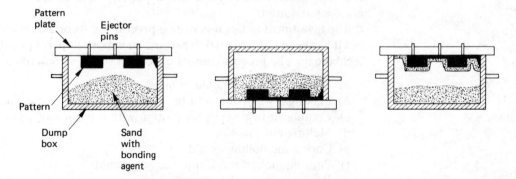

The **shell moulding process** is basically a sand moulding method in which the clay bond present in ordinary foundry sand is replaced by an artificial bonding material. The process is illustrated in fig. 3.23, and consists of three steps:

a) Preparing a thin shell made of a mixture of sand and thermo-setting resin around a heated metal pattern.
b) Separating the shell from the metal pattern.
c) Clamping the shell to a mating shell to form a mould.

Fine silica sand, free from clay, is thoroughly blended with about 5% resin bonding agent (phenol formaldehyde or urea formaldehyde) and placed into a container, called a dump box. The pattern is machined from cast iron, steel or aluminium alloy, and incorporates ejector pins to facilitate stripping of the completed shell. The pattern plate is heated to about 250°C in an oven, then clamped to the top of the dump box which is inverted so that the sand-resin mixture covers the pattern. The pattern quickly becomes coated with a shell of resin-coated sand. Because the resins used are of the thermosetting type, the shell becomes quite hard. The shell thickness depends on the time the sand-resin mixture is in contact with the pattern plate. The dump box is returned to its original position and the surplus, unaffected sand-resin falls back into the box. The pattern plate is removed and the shell is released by the ejector pins. Further hardening of the shell is achieved by a final curing for a few minutes at about 320°C. The two halves of the mould are then joined together by adhesives or mechanically by nuts and bolts or spring clamps. The mould is then ready to receive its charge of molten metal. In the case of large, heavy castings the shell mould may be supported by coarse sand or steel shot held in a container.

Cores may be readily produced by blowing the sand-resin mixture into a chamber instead of using the dump box method.

The main advantage of shell moulding is the high dimensional accuracy of the product. Also the surface finish obtainable is far better than that of sand castings. Shell moulds are not so easily damaged in store and are more readily transported. Working conditions are much cleaner than with ordinary foundry sand and labour costs are lower. Most foundry alloys, including cast iron, plain carbon steels, stainless steel, copper-base alloys and aluminium-base alloys, can be successfully cast into shell moulds. The main disadvantage is the high cost of making the metal patterns due to the high dimensional accuracy required.

The **investment or lost wax casting process** had its origins in ancient history, but it has been rediscovered to become an important part of modern founding technology. The process consists of several stages as described below:

a) Making a master mould in metal.
b) Making a wax pattern of the required article in the metal mould.
c) Coating (investing) of the pattern with a refractory shell.
d) Melting out the wax.
e) Curing the hollow mould.
f) Pouring molten metal into the hot mould.
g) Removing the solid casting.

The master mould (or die) is produced in an easily machinable material such as free-cutting steel or brass. The mould may be of only two parts or it may be a complicated assembly but, in any event, it must allow easy removal of the wax replica. The shrinkage of both metal and wax is allowed for in the design of the die. A large number of castings must be required to justify the high cost of the die.

A low melting point wax is then injected under pressure into the mould cavity and after solidification the wax pattern is removed from the mould. Depending on size, several wax patterns may now be attached to a central runner also made of wax. The complete wax assembly is dipped into a slurry of china clay (fine clay suspended in water) and excess slurry allowed to drain off. The pattern is then given a coating of fine dry fireclay particles which are sprayed on. Several coatings are given to the pattern to build up the final mould thickness to about 1 cm. The investment is allowed to dry in air, and when it has hardened it is heated to about 150°C so that the wax melts and runs out, leaving a mould cavity in the investment material. The refractory mould is then heated to 700–1000°C to remove the last traces of wax and to form bonds between the refractory particles. Molten metal is poured into the moulds while they are still hot, so that the metal is not chilled, but has sufficient fluidity to fill even the thin sections. The molten metal flows into the mould under gravity, but with intricate castings it may be necessary to inject the metal under pressure, e.g. by mounting the moulds in a centrifuge. The castings are usually knocked out of the moulds on vibrating shakers.

Investment casting is especially useful for producing small components in materials which are very difficult to forge and machine, e.g. blades for gas-turbine and jet engines. The dimensional tolerances obtainable are very good and another advantage is that there is no parting line such as appears on castings made in a two-part mould.

The main disadvantage of the process are its high cost and the limitation of component weight to about 2 kg.

Fig. 3.24 Centrifugal casting

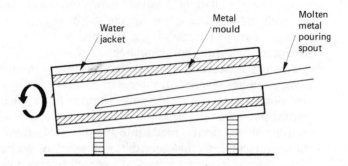

The **centrifugal casting process** is particularly suitable for cylindrical hollow products, e.g. large pipes for carrying water, gas or sewage. The mould, which is tilted slightly off the horizontal (fig. 3.24), consists of a water-cooled metal cylinder which is rapidly rotated about its longitudinal axis. There is also lateral movement between the mould and the entering metal—either by moving the mould or by using a long, retractable pouring *spout* for the molten metal. The centrifugal force forces the metal to the surface of the mould and causes the metal to spread evenly along the length of the mould, so that a hollow cylinder of uniform wall thickness is produced. The relative movement between mould and entering metal is adjusted to give the desired wall thickness. Products up to about 5 m long can be cast. The centrifuging action causes any gases present in the metal to escape to the bore resulting in a sound casting with a uniformly fine-grained outer surface. The method can be used for producing cylindrical shell bearings in a number of different alloy types (e.g. copper-base alloys and lead-tin alloys).

In **die-casting**, the molten metal solidifies in a permanent metal mould with the resulting rapid cooling rate producing castings with finer grain size and hence better mechanical properties than sand castings. Also greater dimensional accuracy, better surface finish as well as higher rates of output are obtainable. However, some alloys which can be readily sand cast do not give satisfactory results when die-cast due to their excessive shrinkage.

Die-casting methods may be classified according to whether the molten metal enters the die under *a*) gravity, *b*) high pressure or *c*) low pressure.

Gravity casting is usually referred to as *permanent mould casting* and the term *die-casting* is reserved for the *pressure die* processes.

In **permanent mould casting**, the charge of molten metal may be poured by hand or may be fed automatically. Moulds are of a simple type, consisting of two halves opening in the vertical plane but are still costly to produce; they are made of cast iron or steel and are provided with risers to feed solidification shrinkage. The mould may be slowly tilted into the vertical position during pouring to avoid splashing, turbulence and possible air entrapment. Any metal cores of complex shape are split to facilitate removal from the finished casting, or alternatively sand cores need to be used.

The process is only suitable for fairly simple shapes without thin sections because the liquid metal freezes rapidly. It is mainly used with alloys which melt below about 1000°C. The whole process may be automated, with the metal being melted in a small, induction, tilting furnace and the mould assembly arranged near the tapping spout.

The main advantage of the process is that a better-quality casting is produced than with sand moulding.

Pressure die-casting involves the injection of molten metal into a mould cavity, and has several advantages over gravity feeding. No riser is needed in the mould, but very small vents (about 0.1 mm thick) are provided at the parting line of the die-halves in order to allow the escape of air. The speed of transfer of the molten metal into the die cavity results in little loss of heat during injection, so that very thin wall sections can be cast and liquid metal can be forced into the recesses of a complex-shaped mould. Rapid solidification under pressure largely eliminates the effects of shrinkage. Again there is little wastage of metal, and the product requires little, if any, machining.

High-pressure die-casting is used for the manufacture of relatively small castings at high production rates. Two types of casting machine are in use:

1   In the *cold chamber machine* the metal is melted in a crucible which is separate from the injection machine: the molten metal is ladled by hand or fed automatically into the injection cylinder (fig. 3.25).
2   In the *hot chamber machine* the metal-holding furnace is incorporated with the injection unit. This type is also called the "gooseneck" machine (fig. 3.26) because of the shape of the injector.

The most widely used machine is the cold chamber type which can produce medium-sized castings in zinc-base alloys (e.g. up to 20 kg in weight) and in aluminium-base alloys. A fairly recent development is the production of small castings in ferrous materials including cast iron, plain carbon steel and stainless steel. The high temperatures involved in the latter application require the use of highly alloyed heat-resisting materials for those parts of the injection unit which come into contact with hot metal.

The molten metal is forced into the die by the piston. After solidification the die is opened and the casting moved away with the platen, from which it is separated by ejector pins.

Fig. 3.25 Cold chamber die-casting machine

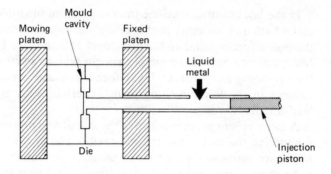

Fig. 3.26 Hot chamber die-casting machine

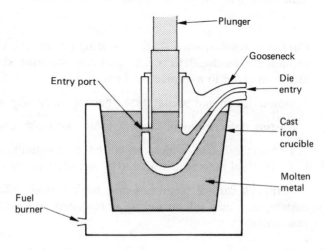

Fig. 3.27 Low-pressure die-casting method

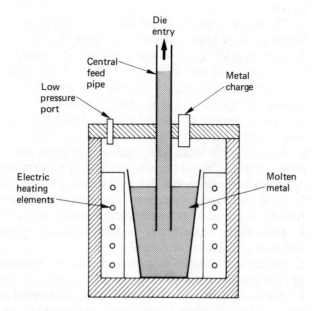

In the hot chamber machine the main part of the injector lies in the molten metal bath and the entry port is below the metal surface. Pressure applied to the piston forces metal up the gooseneck into the die. This type of machine is limited by the amount of metal that can be contained in the injector and by the operating temperature. Therefore the machine is used mainly for small castings in the lower melting point materials, such as zinc-base alloys and lead-tin alloys.

A more recent process involves the use of *low pressure* (about $0.1 \, \text{N}/\text{mm}^2$) for forcing the metal into the die cavity. The metal is melted in an electric resistance furnace, the interior of which is pressurised causing molten metal to be driven up a central feed pipe (fig. 3.27). Fairly large castings, mainly in aluminium-base alloys, are made by this process.

3.8 Casting Defects

Care is needed at each stage of casting production in order to control the incidence of **casting defects**. The quality of the finished casting depends on the attention given to a number of factors including the following:

mould design and preparation, melting, degassing and pouring.

Casting defects may be classified into two main groups:

1   Surface defects which can be detected visually.
2   Internal defects identifiable by non-destructive testing methods.

Although surface blemishes may not adversely affect the properties of a casting, they may still be a cause of rejection, unless they can be removed by machining.

A brief description of various types of **visible defects in sand castings** is given below:

1   *Shrinkage:* appears as unsound areas of shallow depressions in the surface. Shrinkage may cause a decrease in strength and pressure tightness.

2   *Gas porosity:* appears as holes of smooth outline. If these holes are linked to internal pores the pressure tightness will be reduced.

3   *Sand inclusions:* easily identified as particles of sand embedded in the surface. These may cause maching difficulties.

4   *Cracks:* appear as lines of varying depth but, provided they are shallow, may be machined out.

5   *Scabs:* appear as unsightly surface blemishes and are usually found on the cope surfaces. They are usually associated with entrapped sand, and in severe cases cracking may occur due to the differential expansion of the sand.

6   *Cold shuts:* recognised by a wrinkled surface appearance. These may be of variable depth and can significantly weaken the casting.

7   *Cross-joint:* appears as dimensional inaccuracy due to mismatching of top and bottom halves of the mould.

8   *Stickers:* recognised by lumps of metal in a recessed part of the casting. They result from part of the mould face failing to strip from the pattern.

9   *Flash:* appears as thin "plates" of metal formed at joints between mould faces because of a gap leading from the mould cavity.

10 *Warping:* recognised as a movement away from the pattern shape, which occurs during cooling after solidification.
11 *Misrun:* appears as incomplete filling of the mould cavity resulting in lack of reproduction of pattern detail by rounding of corners and thin sections.

Visible defects in die-castings may be listed as follows:
1 *Checking:* appears as surface marks (also called "scores") on the casting caused by imperfections on the surface of a well-used die.
2 *Fins:* appear as thin layers of metal coincident with the parting line of the die when the latter has not closed properly.
3 *Hot tears:* appear as linear cracks usually in thin sections of the larger die castings. Similar cracks may be formed due to the hot-shortness of the alloy itself (e.g. the presence of a minute amount of tin in zinc-base alloys causes hot shortness). The cracks occur due to contraction introducing tensile stresses which the hot metal is unable to withstand.

B.S. 2737 *Terminology of internal defects in castings as revealed by radiography* gives a comprehensive list of **internal** defects with illustrative prints of typical radiographs. Such defects may be broadly classified as follows:
1 *Voids:* a void is a general term used to describe various types of unsoundness ranging from gross shrinkage to microporosity. Voids are caused either by a failure to counteract solidification contraction or by the entrapment of air within the casting.
Various other terms are used to describe certain types of unsoundness, including the following: sponginess, blowholes, pinhole porosoity, microporosity, wormholes and airlock shrinkage.
2 *Cracks:* a crack is a term which describes various types of discontinuity, including hot tears and stress cracks. Cracks may be distinguished by the shape of the path they follow—whether their contour is regular or not.
3 *Cold shuts:* this term refers to a discontinuity which forms when two streams of molten metal meet within the mould cavity and are too cold to fuse properly.
4 *Inclusions:* this term refers to particles of foreign matter such as oxides, sulphides, silicates, dross, sand, slag which become included in the casting.

3.9 Foundry Safety Precautions

A variety of hazards exist in the foundry, including those associated with hot or molten metal, dust, grit, noise and moving machinery. Danger can exist at every stage from the preparation of the mould to the withdrawal of the casting and its fettling. Contact with molten metal can seriously injure and kill, so that great care is needed to avoid making the slightest mistake. It is most important that a high standard of tidiness and order is maintained, and that all equipment is kept in good condition by regular maintenance.
The prevention of accidents involves safeguarding *a*) the work area, *b*) the work method and *c*) the worker.
Employers have a legal responsibility of providing a safe and healthy environment for all employees. However no safety policy is likely to be

successful unless it actively involves all employees. Experience indicates that accidents are more likely to be avoided by training operatives in sound working practices. Therefore, a common approach to industrial safety is the use of "recommended safe working procedures", which are drawn up and agreed by management and union representatives. These procedures usually incorporate information and recommendations from official sources, such as the publications of the Health and Safety Executive. It is important that the procedures are updated as necessary. In addition, it is helpful to have safety rules, some of which may have general scope, while others may be specific to certain work areas in the foundry. Stringent rules should be enforced regarding the wearing of protective clothing appropriate to the work area as indicated below:

Protection needed	Type of protective clothing
Head and neck	Helmets, head shields
Eyes	Goggles or eye shields
Trunk of body	Fire-resistant overalls
Arms and hands	Gauntlets or gloves
Feet and legs	Safety boots with leggings

Operatives must also know the location of first aid posts and fire-fighting applicances, as well as the position of master control switches, so that an emergency shut-down can be carried out if required. These appliances and switches must be readily accessible. There must be sufficient fire extinguishers of the right type located at strategic places throughout the plant and operatives should be instructed in their use. Empty extinguishers must be promptly recharged. After contact with hot metal hand appliances (e.g. tongs) should be marked and set aside so that they will not be accidentally touched. Where very high temperatures are involved, operatives should be protected from heat radiation, e.g. by portable screens. Furnaces and molten metal containers may be insulated by suitable refractory bricks to decrease the heat radiation. All furnaces must be operated in a safe manner to minimise the risk of explosion, fire and spillage or splashing of molten metal. Metal splashing is frequently due to the presence of moisture. Pre-heating is the safest way of ensuring that all surfaces coming in contact with molten metal are perfectly dry.

Fumes may be given off from molten metals and these may be injurious to health. Gas alarms may be used which give warning of escaping gas.

Exercises 3

1 *a*) Make a simple sketch of a melting furnace in which the charge is in contact with both fuel and the products of combustion.
   b) Name four technical factors which influence the choice of melting furnace.

2 Give the essential practical details of two methods of alloying one metal with another.

3 *a*) Make well labelled sketches of three types of steel ingot.
   b) Name two types of ingot defect which arise from different sources.

4 Write a short account of the importance of continuous casting, listing its main advantages and disadvantages.

5 *a*) With the aid of sketches describe the making of a sand casting.
   b) List four properties of a moulding sand needing control.

6 Name four methods of producing castings and describe one of them in detail.

7 *a*) Name two different types of alloy suitable for high-pressure die-casting.
   b) Name two types of machine used for high pressure die-casting.
   c) Make a simple sketch of one of the above-named machines.

8 *a*) Name two main groups into which casting defects may be classified.
   b) Briefly describe four different defects liable to occur in sand castings.

9 State four safety precautions particularly relevant to foundry work.

4 The Working of Metals

4.1 The Importance of Working Processes

The methods by which metallic materials are mechanically shaped into other product forms are called working processes. The very extensive use of metals is due in large measure to their ability to tolerate considerable amounts of permanent deformation without fracture. The products resulting from the working of metals are known as *wrought products*. The processes used to change ingots into wrought forms are called *primary working processes*. Further working by additional methods is often required and these are known as *secondary working processes*. Although the main purpose of working a metal is to produce the required shape, the process may also result in an improvement in the structure and properties of the material. In the production of a finished shape in wrought metal, the size of the starting material and the sequence of the operations must be such that thorough working throughout the cross-section of the material is given. Metal that is worked largely in one direction usually has different properties in different directions relative to the working direction. A large amount of product of a specific shape must generally be required to justify its complete production by mechanical working; otherwise the shape may be machined or fabricated from standard wrought stock.

Depending on the temperature range in which the working is carried out, working processes can be classified as either *cold-working* or *hot-working* operations. As the temperature of a metal is raised, its strength decreases and the metal becomes softer and more plastic. Hence hot working needs less power than cold working and the deformation can be carried out more rapidly. Because the tensile strength is much reduced at high temperatures, hot working usually involves the use of compressive stresses. On the other hand, cold working achieves a high-quality surface finish and is usually used in the final stages of shaping. [The more theoretical aspects of the deformation of metals are dealt with in chapter 5 of *Structure*.]

4.2 Cold Working

Cold working is usually carried out at room temperature and is often the finishing stage in production. The effect of cold working is to distort and elongate the grains in the direction of working (fig. 4.1). The metal becomes harder and stronger as internal stress is increased so that the characteristic ductility is much reduced. The increase in hardness resulting from the plastic deformation caused by cold working is referred to as *work hardening*. As cold working proceeds, the degree of work hardening is increased, the metal loses

Fig. 4.1 Cold working: grains elongated and distorted

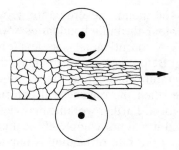

ductility, and the metal requires an increasing applied stress to cause further deformation. A stage is reached when any attempt to cause further deformation will cause fracture of the metal. Therefore, in order to be able to continue the working, the internal stresses must first be removed and ductility restored. This is achieved by an annealing process, which consists of heating the worked metal to above the *recrystallisation temperature* so that new, stress-free, polygonal grains are nucleated within the cold-worked structure. In order to ensure that the whole structure consists of these new grains, adequate time must be allowed for recrystallisation to be complete. The higher the annealing temperature used, the shorter the time required for complete recrystallisation. However, once recrystallisation has occurred, then grain growth takes place, so that the higher the annealing temperature and/or the longer the time, the larger the resulting final grain size.

After recrystallisation is complete, the material is in the fully softened condition and capable of further cold working. However, the need for such inter-stage annealing increases production cost and, in order to minimise this, metals are first hot-worked to near the final shape and then finished by cold working. The material may then be annealed, or it may be left in a cold-worked condition which is hard and strong but of rather low ductility. In between the fully annealed, completely softened condition and the severely cold-worked fully-hard condition, a range of properties is obtainable by suitable combination of cold working and inter-stage annealing. Some metals and alloys in sheet and strip form are available in a number of conditions, e.g. $\frac{1}{4}$, $\frac{1}{2}$ and $\frac{3}{4}$ hard, as well as fully annealed and fully hard.

The advantages of cold-worked products over hot-worked material are:

a) Cleaner, brighter finishes.
b) More accurately controlled dimensions.
c) Increased hardness and tensile strength (but limited ductility).
d) Improved machinability.

4.3 Heating for Hot Working

The temperature to which the material is heated before hot working begins must be carefully controlled for economic reasons as well as to avoid excessive oxidation of the surface and the development of very large grains. The time for which the stock material is kept in the furnace should be the minimum necessary to achieve the appropriate temperature uniformly throughout the mass of metal. Severe grain growth together with the penetration of oxygen along the grain boundaries, forming brittle oxide films, is known as "burning". A "burnt"metal is useless for any service and fit only as scrap for re-melting. Before hot working begins, the mass of metal should be at a uniform temperature throughout the section. If the outside is much hotter than the centre, then the outside will deform more easily than the centre and a split may form. Further working may close this split, but because the surface has oxidised it will not completely seal up. With some materials (e.g. high carbon steels) the rate of heating is important because they may crack if the temperature is increased too rapidly. Such materials should be pre-heated at some intermediate temperature before being placed into a high temperature furnace.

4.4 Hot Working

Hot working is a shaping process carried out at temperatures above the recrystallisation temperature of the material being worked. Deformation and recrystallisation take place at the same time, and no distortion of grains or work hardening occurs. Typical hot-working temperature ranges for a few common metals are given in table 4.1.

Table 4.1 Hot working temperatures

Metal	Melting point °C	Approx. Recrystallisation Temp. °C	Hot-working range °C
Iron	1535	450	1200–900
Copper	1083	200	900–650
Aluminium	660	150	500–350
Zinc	420	20	170–110

Hot working should be completed by the time the temperature of the material has cooled to just above the recrystillisation temperature, so that the finished product will have a fine grain size with good mechanical properties. If the working is finished at a temperature far above the recrystallisation temperature, then the final grain size will be large, resulting in a low-strength material. Hot-worked material is usually cooled rapidly after the final working in order to minimise grain growth. Some materials contain brittle constituents and/or low melting point impurities at grain boundaries and these can be troublesome during hot working. At hot-working temperatures these impurities may melt, leaving the grain boundaries weakened and the grains separated by liquid, so that, on working, the metal crumbles and is said to be "hot-short".

Another difficulty that arises in working operations is that the degree of deformation varies across the section being shaped. A heavy amount of working is more likely to result in a uniform level of deformation throughout the section. Also, during working, the hot metal cools faster on the outside than at the centre so that there is a temperature gradient across the section. The net result of these variations is that the final structure of the wrought product is not uniform and there will be variation in properties.

The major advantages of hot working as compared with cold working are to be found in the lower power consumption and the almost unlimited amount of deformation which is possible. In addition to shaping the metal, hot working may also result in the following benefits:

1 Elimination of the "as-cast" structure (with its coarse, columnar crystals) as well as of coring and the reduction of segregation.
2 Refining of the grain size, provided the finishing temperature is just above the recrystallisation temperature.
3 Welding up of cavities (e.g. blowholes in rimming steel), provided they have clean, unoxidised surfaces.
4 Improvement of mechanical properties (but with some directionality—see section 4.5).

4.5 Directionality in Wrought Material

In cold-worked material (e.g. cold-rolled sheet and strip) the metal grains are aligned and elongated in the direction of working so that a preferred orientation is produced. As a result the mechanical properties of the material may vary considerably depending on the direction of measurement relative to the working direction. The properties are least satisfactory in a direction at right angles to the working direction; the difference in the values of percentage elongation being much more pronounced than is the case with tensile strength. The effect of hot rolling on the mechanical properties of mild steel are shown in table 4.2.

Table 4.2 Effect of hot working on properties of mild steel

Condition	Tensile Strength N/mm^2	% Elongation	Charpy value J
As cast	400	15	10
Hot rolled (in rolling direction)	500	40	50
Hot rolled (at 90° to rolling direction)	450	20	2.5

During hot working, solid non-metallic inclusions present in the metal may be plastic at working temperature so that they may be elongated and form fibres. The elongation and alignment of inclusions give a fibrous appearance to the metal and indicates the direction of flow of the metal, i.e. the working direction.

The main **WORKING PROCESSES** are:

1 Rolling for the manufacture of products such as bars, rods, plate, sheet and strip, as well as shaped sections such as steel girders and joists.
2 Forging for the production of components of fairly simple shape, but with better mechanical properties than castings.

3 Extrusion for the production of solid and hollow sections (e.g. tubes) in both ferrous and non-ferrous materials.
4 Drawing of bar, rod, tube and wire.
5 Pressing and deep drawing of sheet into a variety of components.
6 Miscellaneous operations such as spinning.

4.6 Rolling

Rolling may be carried out as a hot or cold operation. The process involves feeding the metal into the gap between rolls revolving in opposite directions. The compressive forces exerted by the rolls cause the metal to deform and an increase in length is obtained as a result of the reduction in section, the metal leaving the rolls faster than it enters. In the manufacture of flat products plain rolls are used, while grooved rolls are necessary for producing sections. In the whole process of producing a rolled product, several rolling-passes involving different sets of rolls may be needed. Each set of rolls is held in a housing and is referred to as a mill stand. Several mill stands may be used in conjunction to make up the rolling mill.

Hot Rolling

Hot rolling is used in the initial stages of deformation and the various types of rolling mill stand employed are illustrated in fig. 4.2.

A *two-high mill stand* is the simplest type and is widely used for initial rolling of ingots. The mill can usually be reversed, so that the material (stock) can be passed through the rolls in both directions. However, a lot of power is required to stop and start the rolls and this is also time consuming.

A *three-high mill stand* consists of three rolls of equal size vertically above one another. The material being rolled passes between the middle and bottom rolls in one direction, and then a lifting table enables the material to be returned between the middle and top rolls. This is carried out with the mill turning at a constant speed in one direction, so that power and time are saved as compared with a two-high mill.

A *four-high mill stand* has a pair of small work rolls and two larger back-up rolls. Small-diameter work rolls are advantageous in requiring a lower rolling load than larger rolls for a given reduction. Since the rolls for the reduction of wide material must be rather long there is a tendency for the rolls to bend under the rolling load. This is overcome by supporting the small work rolls with more massive back-up rolls. These back-up rolls remain in contact with the work rolls during the rolling so that this type of mill is only used for flat products. The four-high mill can be operated both as reversing and as a non-reversing mill.

The *planetary* mill (fig. 4.3) is a special type of hot rolling mill which was developed to achieve a very large total reduction (about 90%) in a single pass. The mill consists of two back-up rolls surrounded by a number of small work rolls. The work rolls revolve around the back-up rolls which are driven. The action of the work rolls is to perform a great many small reductions in one

Fig. 4.2 Hot rolling mill stands

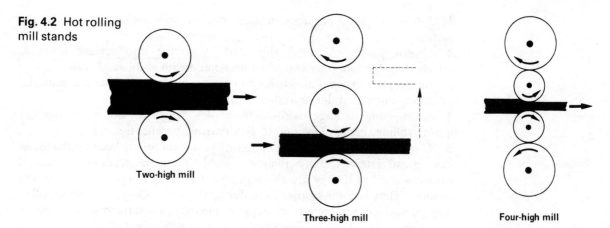

Two-high mill

Three-high mill

Four-high mill

Fig. 4.3 Planetary mill

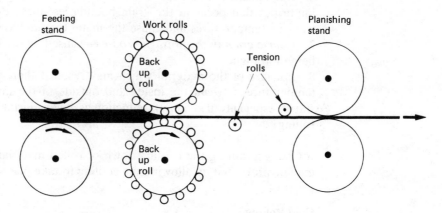

Feeding stand

Work rolls

Back up roll

Tension rolls

Planishing stand

Back up roll

pass, rather like a forging operation, and this cumulative process achieves the very high total reduction at a single pass. A feeding stand forces the material into the planetary rolls and a planishing stand smoothes the rippled strip which emerges. Two small rolls positioned between the planetary rolls and the planishing stand adjust the tension of the strip. The mill works most efficiently on a continuous basis, but can be used on a variety of materials even those with high resistance to deformation.

Defects in hot-rolled products Because rolled products usually have a high surface area to volume ratio, the condition of the surface during all stages of the rolling process is most important. Defects in hot-rolled products can result from flaws introduced during previous stages of production or during the rolling process itself. Occasionally small pieces of refractory or slag may be rolled into the surface. In order to maintain high quality, some surfaces (e.g. surfaces of intermediate products such as ingots, billets) must be conditioned by grinding, chipping or burning with an oxygen lance to remove defects.

Typical defects in hot-rolled products include the following:

1 *Rough or pitted surfaces*—usually due to scale being rolled into the surface.

2 *Slivers*—pieces of metal elongated by rolling into surface tongues, attached to one end only and often associated with embedded scale.

3 *Reeds*—due to *small* blowholes on or near the surface of the cast material becoming elongated during rolling.

4 *Seams*—due to *large* oxidised blowholes, which do not weld together during rolling, being elongated to give narrow, parallel fissures.

5 *Laps*—due to the formation of "flash" in section rolling because the metal has spread laterally more than it should. These projections are called over-fills (or fins if they are thin) and are formed on opposite sides of the product. They become oxidised and during the next rolling pass will be rolled into the surface, leaving a crack between the overfill and the rest of the metal so causing a lap. Laps are usually due to poor roll pass design.

6 *Hot shortness* (see page 104)—due to low melting point constituents or impurities that occur at the grain boundaries; these exist in liquid form at working temperatures and cause the metal to crumble during hot working.

7 *Coarse grain in the finished product*—caused by the finishing temperature being too high.

8 *Fracture* of the material which may result if there is a big difference in temperature between the inside and outside of the material before rolling starts, especially if a heavy reduction is given during the initial stages of rolling.

Cracking during the cooling down of rolled material, especially sections, may be prevented by allowing the cooling to take place in heated chambers.

Cold Rolling

In cold rolling, more complex mills may be used because of the need for accurate control of dimensions. The high rolling loads involved require small work rolls with effective backing up. Mills with four or more back-up rolls are called *cluster mills* (fig. 4.4). As the number of back-up rolls increases, the size of the work-rolls decreases, so that in the Sendzimir cluster mill (fig. 4.5) the work rolls may have a diameter of only 5 cm.

It is common practice to use several mills consecutively. A continuous mill consists of a series of stands in which the stock passes from one stand to the next, and is being rolled simultaneously in several stands. As the stock becomes thinner its length increases so that the speed of rotation of each successive pair of rolls is increased. The consecutive stands may be in tandem as shown in fig. 4.6.

Defects in cold-rolled products The following are typical problems which can arise in cold rolling:

1 *Scratches* due to defective rolls or guides can be a serious quality problem in sheet or strip.

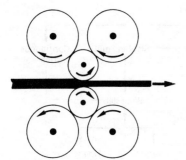

Fig. 4.4 Cluster mill

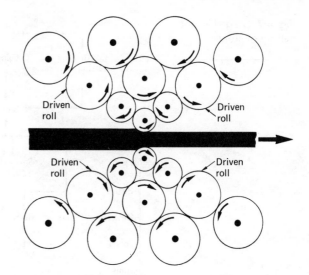

Fig. 4.5 Sendzimir mill

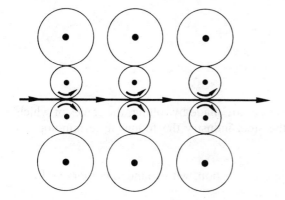

Fig. 4.6 A tandem mill
of three four-high
stands

2 *Cracking due to excessive work hardening.* A controlled amount of work hardening is used to give the required mechanical properties. However, too much working will cause the metal to be in a highly stressed condition and eventually cracking will result.

3 *Cold-shortness* due to a brittle constituent at the grain boundaries.

4 *Variation in product thickness* along the rolling direction due largely to changes in rolling speed or the tension of the material. Variation in either the thickness or hardness of the material entering the rolls will also cause a variation in the thickness of the product as well as roll eccentricity. Sophisticated devices are used to continuously measure the material thickness and adjust the mill operating conditions in order to achieve thickness (gauge) control.

Thickness variation in the width direction is related to the deflection of the rolls. A set of parallel rolls will deflect under load to produce a sheet which is thicker at the centre than at the edges. To compensate for this, it is usual to camber the rolls so that the working surfaces are parallel when the rolls deflect under the load. The rolls on a two-high mill are made a little thicker at the centre than on the outside, and this counteracts the tendency of the rolls to spring open a little at the centre as the stock passes through. In four-high mills the rolls are made slightly thinner at the centre than the outside because when backing-up rolls are used there is very little springing of the work rolls.

Fig. 4.7 Stages in hot rolling of steel

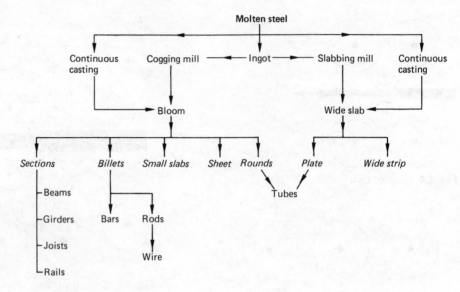

4.7 Rolled Products

Products of various shapes and sizes are made by rolling, the stages of which are indicated in fig. 4.7. In the steel industry the following terminology is conventional:

Bloom: roughly square in cross-section, with rounded corners and larger than 15×15 cm.

Billet: similar to a bloom, but up to 15 cm $\times$ 15 cm in size.

Slab: rectangular in cross-section, with rounded corners and more than 5 cm in thickness.

Bar: straight length with a variety of sections, e.g. flat, round, square, hexagonal.

Rod: round sections smaller than round bars (up to about 10 mm diameter).

Plate: long, wide, flat product, thicker than about 5 mm.

Sheet: similar to plate, but thinner than about 5 mm.

Sections: various shapes of cross-section for construction, e.g. beams, girders, joists, rails, angles.

Blooms, billets and slabs are semi-finished products, i.e. they are an intermediate stage in the production, and will be further processed to give finished products.

4.8 Forging

Forging is usually a hot working process and may be carried out either under a hammer in which the blow has a high velocity, or under a press which exerts a squeezing action. In general, the hammer blows are relatively light, while presses exert a heavy load. Consequently the deformation extends to a much greater depth in press-forged material than in hammer-forged material.

Fig. 4.8 Heavy press forging

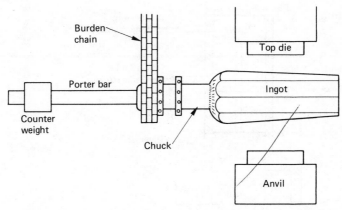

The process may be carried out using dies of simple shape which take successive bites along the length of the stock to achieve the overall reduction. This is known as *open die-forging*. Alternatively, complex shapes can be produced by forging stock between top and bottom dies having the required profile. At the end of the process the dies will completely enclose the formed stock. This is known as *closed die-forging* or *drop forging*.

In *press forging* the starting material is an ingot, and the method may be subdivided into heavy and light press work according to the size of the initial ingot.

In **heavy press forging**, the starting ingot may weigh up to about 200 tonnes and the forging is carried out in a huge hydraulic press in which the metal is squeezed between the anvil and the top die (fig. 4.8). The ingot is cast with a feeder head which allows it to be supported during forging and enables the ingot to be rotated and moved in and out of the press.

The first operation is to fit a chuck and porter bar on to the feeder head so that it can be handled. The ingot is then heated slowly to the required temperature to reduce the risk of cracking; it is then removed from the furnace and lifted by means of a burden chain slung from an overhead crane. The ingot is balanced horizontally by a counterweight fitted on the porter bar then placed between the anvil and the top die to allow forging to begin. Pressing is a slow process and as the steel cools it needs to be reheated several times before forging is complete.

The products of the heavy press include large crankshafts and propellor shafts for ships, rolls for rolling mills and boiler drums.

Light press forging is similar to heavy press forging except that the starting ingot is about 5 to 20 tonnes in weight and manipulators are used instead of an overhead crane and burden chain; the forging is carried out at a faster rate than in the heavy press.

Hammer forging is used to shape ingots weighing less than about 5 tonnes. Lengths of hot rolled bar may also be forged by hammer. The material is given a number of sharp blows and the work is carried out much faster than under a press, but the effects of the work do not penetrate as deeply as in press forging. Massive foundations are required because the shock of the hammer blows must be absorbed.

Fig. 4.9 Double-acting steam hammer

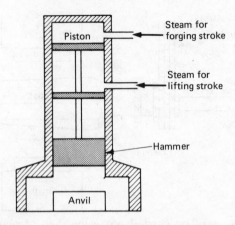

In the double-acting steam hammer (fig. 4.9) the applied force can be varied from a light tap to a heavy blow with the full weight of the tup (top striking face) and the steam pressure behind it. The anvil is usually about twenty times the weight of the hammer in order to absorb the shock. An example of the use of hammer forging is the production of railway carriage axles from round steel bar.

Fig. 4.10 Drop forging

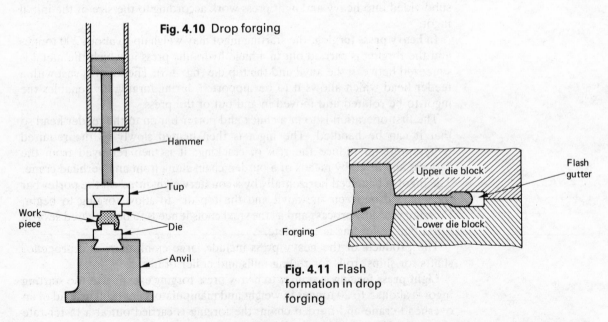

Fig. 4.11 Flash formation in drop forging

Drop forging is concerned with the production of relatively large numbers of forgings from one die block. One half of the die block is attached to the anvil and the other to the tup. The hot metal is forced into the impression formed by the two halves of the die (fig. 4.10). To ensure complete filling of the die, excess stock metal must be used and the dies must have a flash gutter to collect this excess in order to allow the dies to close and give the correct dimensions (fig. 4.11).

In addition to the steam hammer, other types of hammer are used for drop forging including the pneumatic hammer (operated by compressed air) and drop-stamp (in which the tup falls under its own weight).

If the component is of simple shape (e.g. gear blanks) a few blows by the hammer may be enough to complete the forging operation. With a complicated shape, however, the forging will require to be made in stages using more than one pair of dies. Forging must be carried out as quickly as possible because the components cool quickly. The number of forgings produced from one pair of dies may run into thousands with the weight of a single forging varying from a few grammes to several kilogrammes. The accuracy of the finished dimensions should be such that machining is reduced to a minimum and in some cases it is possible to put components into service in the as-forged condition.

The products of drop-forging are numerous and include many components for the motor-car and aircraft industries, e.g. crankshafts, connecting rods, gears, turbine blades.

Defects in forged products   It is most important that the material to be forged be of high quality, free from surface flaws of all kinds. For example, steel for forging must have been completely deoxidised (i.e. killed) before teeming, and it should have a low content of non-metallic inclusion, as well as being free from significant segregation. The heating of the metal preparatory to forging must be controlled. Too high a temperature or too long a soaking time causes severe surface scaling and pieces of scale may be driven into the surface of the component during forging. The following defects may also arise during forging:

1   In closed die-forging the metal may not fill the die properly, resulting in a product of unsatisfactory shape. This may be due to forging at too low a temperature or simply due to insufficient stock.
2   Excessive reduction in thickness during forging may result in flash-line cracks.
3   The upper and lower dies may be out of alignment resulting in mismatched forgings.
4   Unsatisfactory flow of metal may cause two surfaces to fold against each other as the die closes causing a forging lap.
5   If the deformation applied does not penetrate much below the surface, then the interior may still have a coarse structure, so that there will be a variation in properties throughout the forging.
6   Working at too low a temperature may cause surface cracking.
7   Working at too high a temperature may aggravate any tendency towards "hot-shortness".
8   Finishing at too high a temperature may result in a product with a coarse grain size.

4.9 Extrusion

In **extrusion** the metal is squeezed through a die in similar fashion to toothpaste being forced out of a tube. Since large forces are required, the process is usually carried out as a hot-working operation, although cold extrusion is sometimes possible. Because the deformation is achieved entirely by compressive forces, it is possible to extrude relatively brittle alloys. Extrusion is a very important process for the production of bars, rods, sections and tubes in non-ferrous metals and alloys. Large reductions may be achieved by extrusion and this is normally quoted in terms of an extrusion ratio R where

$$R = \text{Original cross-sectional area} : \text{Cross-sectional area of product}$$

Thus an extrusion ratio of 10:1 is equivalent to 90% reduction.

Extrusion processes may be classified as follows:

1   Direct extrusion
2   Indirect (or Inverted) extrusion
3   Hydrostatic extrusion
4   Impact extrusion

Direct Extrusion

A cast metal billet is heated to the required temperature and placed in the container of the extrusion press (fig. 4.12). The hydraulic ram then applies pressure to the billet causing the metal to be forced through the orifice. The container and the die are fixed and therefore remain stationary during the process. Consequently the billet moves relative to the container so that friction arises, causing a peculiar flow of the metal during extrusion inasmuch as the centre of the billet moves forward faster than the outside.

Fig. 4.12 Direct extrusion, showing typical extruded sections

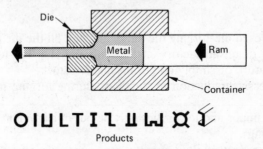

Products

Hollow sections, including tubes, may be produced by forcing a bored billet through a die using a mandrel (fig. 4.13). The billet moves through the die as pressure is exerted and the section is formed in the space between the die and the mandrel. The extrusion may also be carried out by piercing the hot billet and extruding in one operation. Tubes can be extruded cold as well as hot.

Direct extrusion defects   In addition to flaws originating from the cast metal (e.g. gas unsoundness, oxide inclusions), products of the direct extrusion process may suffer from the following defects.

Fig. 4.13 Extrusion of hollow section

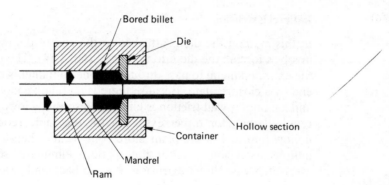

1 *Surface cracking* may occur if extrusion is carried out too rapidly or at too high a temperature, because in these circumstances lubrication may be inadequate and the metal sticks to the dies. The problem is often associated with hot shortness.

2 *Variations in structure and properties* exist because of the differing amount of deformation suffered by different zones. The leading end receives only slight working so that the structure is substantially as cast. Deformation is greater in the outer zones of an extruded bar than in the centre; variation is greatest when the extrusion ratio is low, since the centre then only receives light deformation. Hence above a certain degree of reduction the disparity should get less. The heavily worked outer zone, although extruded above the recrystallisation temperature, may nevertheless be in a cold-worked condition. This is because the rate of working is so great that the rate of crystallisation is much less than the working rate and the material does not stay at the recrystallisation temperature long enough after the operation to allow the cold-worked zones to recrystallise. Any zone with a small residual amount of cold work may develop large grains on subsequent reheating above the recrystallisation temperature.

3 *The back-end defect* (or "extrusion defect") arises from the peculiar flow of the metal during direct extrusion. The outer skin of the metal and associated oxide eventually moves inwards and forwards resulting in the presence of oxide internally in the product. Sometimes a complete ring of oxide film may be formed with the core of the product being completely detached from the outside.

The back-end defect may be overcome in one of the following ways:

a) Allow a back-end discard of 15–20%.
b) Arrange for the oxide layer on the billet surface to be removed in the container (thus leaving a "skull behind") and discarding 7–10% at the end.
c) Machine the surface of the cast billet and then preheat it in a controlled atmosphere.
d) Reduce the friction between billet and container as in the indirect extrusion method (see below).

Indirect Extrusion

In this method the die is attached to the end of the ram so that the process involves forcing the die into the metal (fig. 4.14). The ram is hollow to allow the extruded metal to pass through it. The container is fixed and closed at one end by a closure plate. Because there is no relative movement between billet and container, wall friction is eliminated so that a lower extrusion pressure is required than for direct extrusion for the same reduction. However, the hollow ram makes the plant more complicated than for direct extrusion. In indirect extrusion a better flow pattern eliminates some of the problems associated with direct extrusion, e.g. no back-end defect occurs.

Hydrostatic Extrusion

In this process a fluid is used to transmit the pressure from the ram to the billet. Various fluids have been used, the most successful being mineral oil with an addition of 10% molybdenum disulphide. Direct contact between the ram and billet is eliminated; die lubrication is achieved by the fluid and lower pressures are required. Fig. 4.15 shows the principle of the process. The techniques enable materials with limited ductility, such as high speed steel, to be extruded and if the extruded product is allowed to emerge into a pressurised chamber, then cast iron, tool steels, nickel and titanium alloys can be cold extruded.

Extruded Products

Rods, hexagonal bars and a large variety of sections (including hollow sections) are extruded in many non-ferrous alloys e.g. brass, aluminium-base alloys and nickel-base alloys. The extrusion of steel is also becoming increasingly important. Some sections which may be hot-rolled can also be extruded so that the two processes may be in competition.

Impact Extrusion

This is usually a cold deformation process in which a punch descends swiftly on to a disc-shaped blank positioned in a die, causing the metal to flow upwards and around the punch (fig. 4.16). Collapsible tubes in aluminium, tin, lead and their alloys can be made by this process.

4.10 Cold Drawing   Cold drawing involves reducing the cross-section and increasing the length by pulling the metal through a die at room temperature. The material being drawn should possess high ductility and a reasonably good tensile strength. Common products of cold drawing include bright bars, rods, tubes and wire. The starting material is either hot-rolled stock (ferrous) or extruded material (non-ferrous). A patenting heat treatment process similar to austempering [as described in *Structure*, page 128] is often used on high carbon steel rod or base

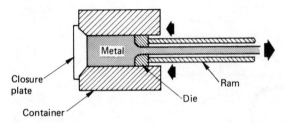

Fig. 4.14 Indirect extrusion

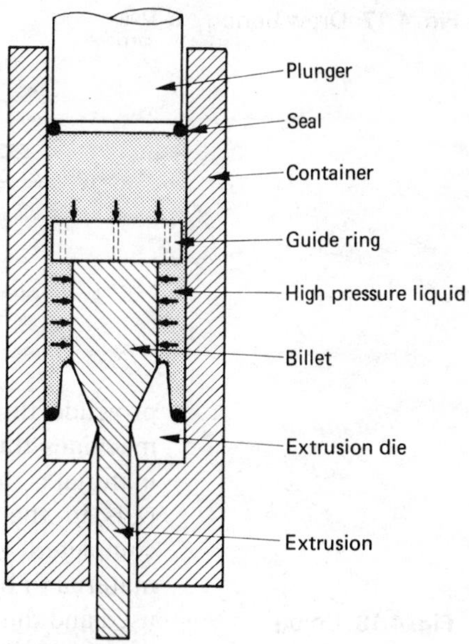

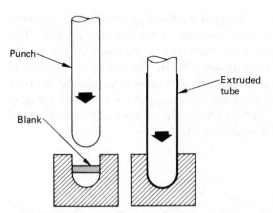

Fig. 4.16 Impact extrusion

Fig. 4.15 Hydrostatic extrusion

wire in order to obtain a suitable combination of strength and ductility. The transformation temperature is chosen to produce a fine pearlite structure.

The surface of the material must be clean and oxide is usually removed by pickling. For example, steel may be pickled by immersion in 3–10% H_2SO_4 at 70°C with added restrainer to prevent attack on the basis metal. After pickling, the material is washed in water. Steels may be drawn either wet or dry. In dry drawing, the pickled rod is dipped into a hot lime solution and drawn using a hard soap as lubricant. For wet drawing, which is more commonly used for fine wires, the starting wire is either phosphated or given a thin coating of copper by dipping in an inhibited solution of copper sulphate. The usual lubricant is then an emulsion of fats and soap in water.

Cold Drawing of Bars, Rods and Tubes

In these cases, cold drawing is carried out in order to to give a bright smooth finished product with closely controlled dimensions. The starting material is only about 1 to 2 mm oversize, so that the amount of reduction is very small. The drawing is carried out on a draw-bench (fig. 4.17) which must be long enough to take the length of the product required (usually up to about 12 m). The draw-bench consists of a rigidly held die, and a "dog" which grips the end of the bar and pulls it through the die. This dog is mounted on a carriage which runs on rails and is hooked to an endless chain driven by an electric motor. The front end of the material is reduced in section to enable it to

Fig. 4.17 Draw bench

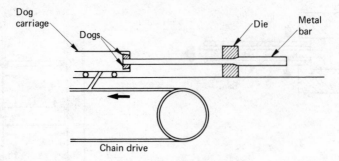

protrude through the die and be gripped by the dog. In modern drawing machines this "pointing" or "tagging" is part of the drawing process. The material is drawn through the die as the carriage and the dog travel down the rails to the far end of the draw bench. When the operation has been completed, the dog is released and the carriage returns to the die for the next material to be drawn. The usual lubricant is grease, but sometimes dry soap is used and this gives a rather dull surface finish. The only waste material in cold drawing is the tagged end which is discarded.

Fig. 4.18 Cold drawing of tube

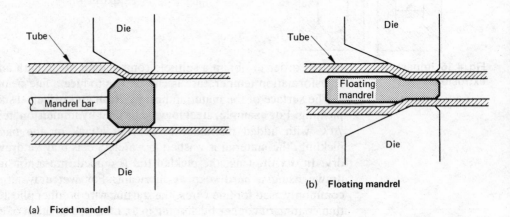

(a) **Fixed mandrel**

(b) **Floating mandrel**

In the production of drawn tube, internal support is needed and this requires the use of a mandrel or plug. For large tubes the mandrel is attached to the end of a long rod which is anchored to the opposite end of the draw bench (fig. 4.18a). The tube to be drawn is first threaded over the mandrel and rod. The rod is then attached to the end of the draw bench and the tube drawn through the space between the mandrel and the die.

For tubes with a diameter less than about 5 mm the *fixed mandrel* method described above cannot be used because the mandrel would have to be very thin to enable it to pass up the tube. The use of a *floating mandrel* (fig. 4.18b) enables very fine tubes to be cold drawn. The mandrel is so designed that it adjusts itself to the correct position for drawing.

Fig. 4.19 Wire-drawing die

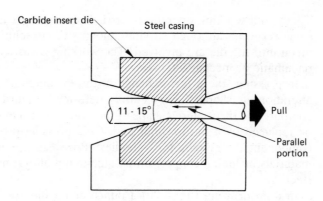

Carbide insert die
Steel casing
11 - 15°
Pull
Parallel portion

Wire Drawing

A very important application of cold drawing is in the production of wire. The starting material is hot worked rod about 5 to 20 mm in diameter. In general, the smallest diameter which will give the required mechanical properties in the finished product is used as starting material. After a certain amount of drawing through the die (fig. 4.19) the material will become so brittle that further working without cracking is impossible. The brittleness is removed by annealing which is carried out in a controlled atmosphere to avoid oxidation. More than one interstage annealing may be necessary during the whole drawing process. As mentioned previously, high carbon steels used for rope or springs are given a heat treatment called patenting and this may be used before, during or after the drawing sequence.

Defects in cold drawn products It is most important that the hot-worked starting material is thoroughly cleaned and is free from surface debris of all kinds; it must have a satisfactory surface before cold drawing begins otherwise surface soil will become embedded into the cold drawn product. The following *surface* defects may arise:

1 *Longitudinal scratches* caused by a scored die, poor lubrication or by abrasive particles being drawn into the die along with the stock material.
2 *Slivers* resulting from swarf being drawn into the surface of the material.
3 *Long fissures* which originate in ingot defects (e.g. blowholes).

Internal cracks may be due to defects present in the hot-worked starting material (e.g. pipe, seams), or rupturing of the centre of the products due to over drawing, this being known as "cupping".

Cold drawing introduces internal stress which may give rise to cracking in certain corrosive environments.

4.11 Pressing, Deep Drawing, and Stretch Forming

These cold working operations carried out on sheet material are similar in some respects, although varying amounts of stretching and drawing of the metal into the die are involved. The working is carried out using hydraulic, pneumatic or mechanical presses.

In **pressing**, there is no change in the thickness of the material, only an alteration in shape. The metal is deformed against two dies which are contoured to the required shape of the component. However, the top die may be replaced by a hard rubber pad. Cold pressing produces various shaped components, e.g. motor car bodies. Pressings are made in a variety of materials including copper, brass, aluminium alloys, mild steel and stainless steel.

In **deep drawing** (fig. 4.20) a punch and a die are used to form cup-like shapes. The metal blank is placed on a die and the metal deformed by a punch and drawn into shape. If more than one drawing operation is involved then intermediate annealing may be necessary, care being taken to prevent surface oxidation by using a controlled atmosphere; otherwise pickling is necessary. Typical components produced by deep drawing include shell cases, milk churns and bath tubs. Only ductile materials are suitable for deep drawing, e.g. 70/30 brass, cupro-nickel alloys and some aluminium alloys. Alloys not ductile enough for deep drawing may be quite satisfactory for pressing operations.

The grain size of the starting material should be uniform, but not too fine, otherwise working may be difficult, nor too coarse or else a roughened surface results ("orange peel" effect).

Defects in pressed and deep drawn products

Defects and characteristics	Causes
Draw marks—vertical score lines on the outside.	Rough draw surface, foreign matter, inadequate lubrication.
Wrinkles on flange.	Insufficient pressure on blank holder, die and blank faces not parallel.
Burnished areas on wall.	Insufficient clearance between die and punch.
Mis-strike—uneven shell height or flange width.	Misalignment of punch and blank.
Earing—high points on top of shell.	Variable physical properties of metal blank (anisotropy).
Orange peel effect—rough appearance.	Grain size of blank too large.

4.12 Spinning and Flow-turning

Spinning is one of the oldest methods of shaping sheet material to produce a hollow shape. The process involves the application of pressure, using a hand tool, to a rapidly rotating circular metal blank so forcing the metal to assume the shape of a former which is rotating with it (fig. 4.21). The former has the internal shape of the finished component. The hand-tool and the former are often made of hardwood. Formers may be solid, but if the finished product is

Fig. 4.20 Deep
drawing

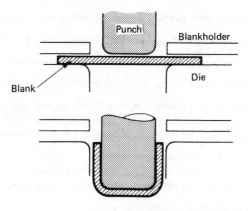

Fig. 4.21 Spinning

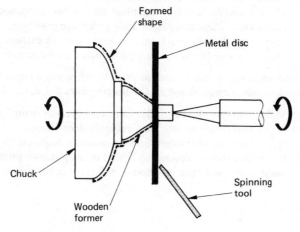

of re-entrant shape, then the former must be segmented to enable easy
withdrawal from the component. The deformation is achieved by combined
bending and stretching.

Although the required equipment is simple, considerable experience is
needed for successful spinning—especially in judging when the metal has
become so hard that annealing is necessary. Conical-shaped components may
be difficult to produce by deep drawing but relatively easy to make by
spinning. Some articles of re-entrant shape can only be produced by spinning.
Typical components made by this process include domestic hollow ware,
dairy utensils and reflectors. Spinning is only suitable for the production of
relatively small numbers of components, and for larger-scale production deep
drawing may be suitable or flow turning used.

Flow-turning is, in effect, a mechanical version of spinning and a high rate
of production is possible with control over dimensional accuracy. In this
process, thick material is plastically deformed by pressure-rolling so that the
metal flows in the direction of travel of the roller. The wall thickness of the
finished component is much less than that of the original blank. Typical
articles produced by this method are aluminium alloy milk churns and
stainless steel dairy vessels.

Exercises 4

1 Distinguish between a primary and a secondary working process.

2 Name four advantages of cold-worked products over hot-worked material.

3 Name four effects of hot working on an as-cast structure.

4 Explain what is meant by the term "directionality" as applied to worked material.

5 *a*) Make sketches to illustrate three types of rolling mill stands.
   b) State the main advantages of a "planetary" rolling mill.
   c) Briefly describe four different defects in hot-rolled products.

6 *a*) Name four types of forging process.
   b) State two advantages of forging over other hot-working processes.
   c) Briefly describe four different defects which may appear in a forging.

7 *a*) Name four types of extrusion process.
   b) Make a simple sketch of the plant used in one extrusion process.
   c) Name two different types of alloy which are commonly extruded.
   d) Name four defects which may be present in an extruded product.
   e) State how one of the mentioned defects may be minimised.

8 With reference to the cold-drawing of tubes, make a simple sketch to illustrate *a*) the fixed mandrel method and *b*) the floating mandrel method.

9 *a*) Make a simple sketch to illustrate the "deep-drawing" process.
   b) Name two different articles commonly deep-drawn.
   c) State the main characteristics of a material required for deep drawing.
   d) Name two defects commonly found in deep-drawn products.
   e) State how one of the mentioned defects may be minimised.

5 Pyrometry and Heat Transfer

5.1 Temperature Measurement

The word "pyrometry" is derived from the Greek "pyr" meaning fire and "metron" a measure. The accurate measurement and control of temperature is of great importance in many metallurgical operations (e.g. hot working, heat treatment) in order to obtain optimum properties from materials and achieve efficient utilisation of energy.

The *absolute scale of temperature* or *Kelvin scale* is derived from the Second Law of Thermodynamics as applied to an ideal gas. This law (the Gas Law) states that:

Pressure of gas × Volume of gas = Constant × Temperature of gas

and is only true for an ideal gas. While none of the gases are perfect in this sense, hydrogen and nitrogen can be used in practice because their known deviation from perfection allow suitable corrections to be made in subsequent calculations.

Both theoretical considerations and practical experience show that it is not possible to reach a temperature lower than a certain minimum, and Lord Kelvin proposed that this minimum temperature be known as the *absolute zero of temperature*. Also, a unique temperature, known as the *triple point of water*, where ice, water and water vapour are all in equilibrium, was adopted as the "fixed point" by which the Kelvin scale is defined: the definition being that there are 273.16 K between the triple point temperature and absolute zero. The triple point temperature may be used for calibration purposes, but in practice the temperature of melting ice, 273.15 K at standard atmospheric pressure, may be used with sufficient accuracy.

The Kelvin scale is not very convenient and the International Temperature Scale is usually used in industry. This is similar to the Kelvin scale but uses the degree Celsius, °C, defined as follows:

Celsius temperature (°C) = Kelvin temperature (K) − 273.15

The temperature interval is the same on both scales, i.e. 1°C = 1K.

In any calculation involving the gas law, absolute temperatures must be used and, for practical purposes, the relationship

K = °C + 273

may be used.

The Constant Volume Gas Thermometer, consisting essentially of a bulb containing hydrogen or nitrogen, measures temperature by application of the Gas Law. As the temperature rises, the gas attempts to expand, but is prevented from doing so and, consequently, the pressure increases. This increase in pressure is measured and since it is directly proportional to the rise in temperature, the latter can be measured.

Though it is a fundamental measuring instrument, the gas thermometer is cumbersome, fragile and unsuitable for the routine measurement of temperature. However, the instrument was used to determine the freezing points of many pure metals and chemical compounds up to about 1500°C and these are used as standards against which a number of secondary pyrometers can be conveniently calibrated. These instruments do not give absolute temperature readings but are calibrated by using fixed points and the results are referred to as being on the *International Temperature Scale*. This is based on six primary points and a number of secondary fixed points. The *primary points* are in °C:

The boiling point of oxygen	−182.97
The freezing point of water	0
The boiling point of water	100
The boiling point of sulphur	444.6
The freezing point of silver	960.8
The freezing point of gold	1063.0

The *secondary points* include the freezing points (°C) of the following substances:

Mercury	−38.89
Tin	231.9
Lead	327.3
Zinc	419.5
Antimony	630.5
Aluminium	660.1
Sodium chloride	801
Copper	1083
Nickel	1453
Platinum	1769
Rhodium	1960
Iridium	2443
Tungsten	3380

5.2 Temperature-measuring Devices

A change in temperature causes changes in the physical properties of materials (e.g. changes in dimensions) and these changes may be used in temperature-measuring devices. Also when objects are sufficiently hot they emit energy as radiation, the detection of which may form a basis for measuring their temperature. Thermometers may be used for temperatures up to red heat, but above this the instruments are usually referred to as *pyrometers*.

Temperature-measuring devices may be divided into *contact* and *non-contact* types or alternatively they may be classified as either *electrical* or *non-electrical* in operation. The specific choice of pyrometer will depend on the temperature to be measured, the accuracy required, the need for recording, the cost, and speed of response needed.

5.3 Contact-type Devices

Such devices rely on physical contact between the temperature sensor and the object whose temperature is being measured. This contact ensures that the sensor and the object are at the same temperature, but limits the use of this type to about 600°C. Contact-type devices include the following.

Paints and crayons which change in colour at a fixed temperature are cheap and especially useful for indicating variation of temperature over the surface of a component, e.g. during local heating prior to welding or in stress-relieving.

Paints are available for use up to 800°C and are claimed to be accurate to within 1% of the measured temperature. *Tempilstik* crayons consist of sticks of salt mixtures which melt at 25°C intervals over the range 80–1100°C. The component is marked with the crayon, which melts at the stated temperature. The accuracy is said to be within 10–20°C of the actual temperature. *Pyrometric Seger* cones are triangular pyramids about 60 mm high, made from mixtures of magnesia, quartz, lime, iron oxide, kaolin, feldspar and boric acid; the composition is so adjusted that the cone bends over and finally the apex touches the base at a fixed temperature (fig. 5.1). Temperatures between 600°C and 2000°C can be indicated by a series of 61 cones, with an accuracy of 25–40°C. The main disadvantages of the use of the Seger cones are that the temperature measurement is affected by the rate of heating and the type of furnace atmosphere.

Fig. 5.1 Seger cones

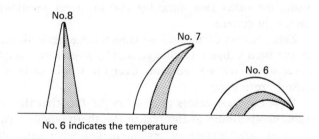

No. 6 indicates the temperature

Expansion thermometers make use of the change in volume of fluids with change in temperature. The simplest type is the *mercury-in-glass thermometer*, but this can only be used up to about 300°C. Such thermometers are used for measuring liquid temperatures, although the liquid should be stirred. Mercury-in-glass thermometers are relatively cheap, reliable, easily portable and do not need additional equipment. The main disadvantage arises from the fragility of glass, and also surface temperatures cannot be measured. The useful range of the thermometer can be extended up to 600°C by filling the capillary tube above the mercury with nitrogen under pressure, which raises the boiling point of the mercury. To increase the robustness of the instrument, the glass bulb is replaced by a steel container and the mercury expands not into a graduated tube but into a Bourdon gauge, the pointer of which indicates the amount of expansion and hence the temperature (fig. 5.2). The

Fig. 5.2 Mercury-in-steel thermometer

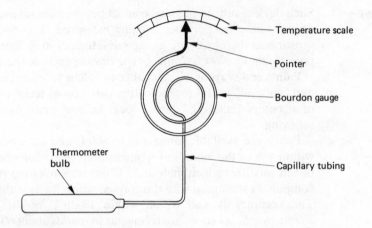

two parts are connected by a tube of very small bore so that the gauge may be placed some distance from the bulb. Thus this instrument may be used in situations where a liquid-in-glass thermometer would be unsuitable.

The liquid-in-glass thermometer is subject to four main **sources of error**, apart from any faults in manufacture and calibration. Two of these, zero change and stem exposure error, are peculiar to this type of thermometer while the other two, time lag and incorrect positioning, apply to thermometers in general.

Zero changes occur because glass is not a perfectly stable solid. The volume of the bulb is liable to change slightly over a period of time, introducing an error which will be the same over the whole scale. Errors of this kind are called "zero errors".

Stem exposure errors arise from the fact that the bulb may not be at the same temperature as the rest of the thermometer. Thermometers are often marked "total immersion" meaning that, when calibrated, the whole thermometer was immersed in a liquid, so that all parts were at the same temperature. If, as often happens in normal use, only the bulb is at the temperature to be measured and most of the stem is at a lower temperature, the liquid column is, in effect, shortened because it is cooler than when the thermometer was calibrated. A low reading will therefore result.

There is a *time lag* in the response of any kind of thermometer. In the case of a liquid-in-glass thermometer, this lag is considerable and when a thermometer is placed in a hot liquid it is some time before its reading becomes steady. This is because heat must be transferred through the glass of the bulb (which is of low thermal conductivity) to the thermometer liquid before the latter reaches the temperature of its surroundings. A thermometer of this type is not capable of giving the true value of a fluctuating temperature.

Incorrect positioning of thermometers is a common source of error. The reading of a thermometer is the temperature of its bulb and this may not be the temperature of the object under investigation. For example, a thermo-

Fig. 5.3 Bimetallic strip

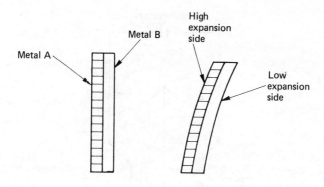

meter in a pipe-line of fluid must project well into the pipe to give a reliable reading.

Another type of expansion thermometer makes use of a *bimetallic strip*. When the temperature is raised, one metal expands more than the other causing the strip to bend; the two metals are known as the high and low expansion sides of the strip (fig. 5.3). In such bimetallic thermometers the strip is usually formed into a helix so that the bending effect is converted into a rotation of one end relative to the other. One end is fixed and the other attached to a pointer. The temperature scale is linear and a temperature range of 0–400°C may be covered. Bimetallic strips are compact and robust but also more expensive than other types of expansion thermometers.

Electrical Resistance Thermometers

For nearly all metallic materials, a rise in temperature causes an increase in electrical resistivity. Over an appropriate range of temperature, this increase is directly proportional to the increase in temperature. An electrical resistance thermometer makes use of this effect and consists of a small coil of wire together with an electrical circuit which measures the change in its resistance. The metal usually used is platinum in the form of wire about 0.1 mm diameter and having a resistance of about 100 ohms. The wire is wound on a mica former which is enclosed in a thin sheath of glass, silica or stainless steel to protect it from the environment.

Platinum, although expensive, is usually preferred to other metals for the following reasons:

a) A simple relationship exists between temperature and resistance over the range −200 to +1000°C.
b) It has high resistance to oxidation and corrosion.
c) It is available with high purity.
d) It has a high melting point.

Fig. 5.4 Wheatstone bridge circuit for platinum resistance thermometer

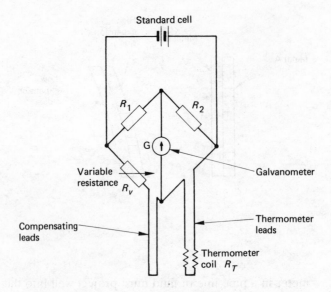

Measurement of the change in resistance is made by the circuit known as a *Wheatstone bridge*, the thermometer coil being made one arm of the bridge as shown in fig. 5.4. The two arms R_1 and R_2 are fixed resistances, while R_v is a variable resistance. A galvanometer G is placed across the bridge, which is connected to a standard cell or battery.

The resistance R_v may be adjusted until G shows no current so that

$$R_1/R_2 = R_v/R_T$$

This method is accurate but time-consuming so that, for industrial use, R_v is left unaltered and changes in R_T cause an out-of-balance current to flow through G, which can be calibrated so as to give a direct reading of temperature.

Although the platinum resistance thermometer has the highest accuracy of all pyrometers over the range −200 to +1000°C it is little used industrially because it is bulky and fragile, and also due to the high cost of platinum. It is, therefore, mainly used as a master instrument in the calibration of other instruments.

Thermistors are similar to resistance thermometers, but use a thermally-sensitive semiconductor instead of a coil of wire. The semiconductor is in the form of a small bead, made of metal oxides, the resistance of which decreases with rise in temperature. The relationship between resistance and temperature is not linear as with metals but, in fact, exponential. However, the temperature coefficient of resistance of the metal oxides is about ten times that of metals, so that thermistors can detect very small temperature changes. The change in resistance may be measured with a Wheatstone bridge circuit as with resistance thermometers. Thermistors may be made small in size and are suitable for the temperature range 0–600°C.

Thermocouples (or thermo-electric pyrometers) consist of two different metallic wires joined at one end. The joined end is placed close to the heat source and is called the hot junction, while the opposite ends of the wires

Fig. 5.5 Typical
thermocouple circuit

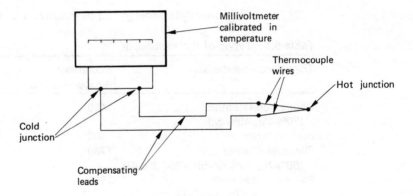

form the cold junction and the circuit is completed with an indicating
instrument (fig. 5.5). Because the junctions are at different temperatures and
the wires are of different materials, an e.m.f. is set up and a current flows in
the circuit. To form a thermocouple, the wires are usually joined by welding,
but the method of joining and the sizes of the wires have no effect on the
e.m.f. generated. The magnitude of the e.m.f. depends on the temperature
difference between the junctions and the wire materials used. Usually, the
greater the temperature difference the larger the resulting e.m.f. The e.m.f.
can be measured by a millivoltmeter which may be calibrated directly in
degrees Celsius.

The e.m.f. generated in a thermocouple circuit results from two effects,
namely the Seebeck effect and the Thomson effect. In the *Seebeck effect* an
e.m.f. is generated between two different metals placed in contact, while the
Thomson effect governs the production of an e.m.f. between the ends of a
homogeneous conductor when one end is heated.

It is seen, therefore, that if one junction is held at a uniform temperature
(the cold junction) then the e.m.f. developed can be used to determine the
temperature of the other junction (the hot junction). This arrangement of
conductors is called a "thermocouple" and is the basis of the most widely used
industrial type of pyrometer. The complete pyrometer consists of the
following parts:

a) Two dissimilar conductors (in wire form, joined at one end) which will
produce an e.m.f. large enough to be measured, and able to function at the
highest temperature required without melting or excessive oxidation.
b) Refractory insulation to prevent the wires from touching, except at the hot
junction; one or both wires are covered with silica or fire-clay capillary
tubing.
c) A sheath to protect the couple from injurious gases or corrosive slags.
d) An instrument for measuring the e.m.f. generated (e.g. a millivoltmeter
or a potentiometer).
e) A means of controlling the temperature of the cold junction.

Types of thermocouple widely used in industry are shown in table 5.1.

Table 5.1 Types of thermocouple

Thermocouple metals	Maximum temperature °C	Millivolts per 100°C (approx)
Copper–constantan (60% Cu, 40% Ni)	400	4
Iron–constantan	800	5
Chromel–alumel (90% Ni, 10% Cr–98% Ni, 2% Al)	1200	4
Platinum–platinum +10% rhodium	1600	1

Fig. 5.6 E.M.F.—temperature relationship for various thermocouples

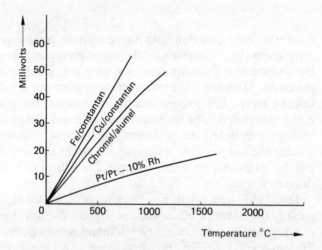

The metals used may be classified as base metals (e.g. Fe, Cu, Ni and their alloys) and noble metals (e.g. Pt). Base metal thermocouples may be used up to about 1200°C with an accuracy within 2°C, while the expensive, noble metal type are suitable up to 1600°C with an accuracy within 3°C at the higher temperatures. The e.m.f. developed by base metal couples is about four times that of the platinum couple (fig. 5.6). Also, because of the lower cost, thicker, stronger base-metal wires can be used resulting in lower circuit resistance and improved life of the couple. However, the wire diameter must not be too large otherwise heat would be readily conducted away from the hot junction. It should be noted, however, that noble metals are available in a purer, more homogeneous form and noble metal couples have longer life at the higher temperatures than base metal couples. The smaller the mass of metal in the hot junction, the greater the sensitivity of the couple to change in temperature. As an industrial instrument the thermocouple has the advantage that the hot junction is of simple and robust construction, and is easily replaced if damaged. For the measurement of molten-metal temperatures the wires can be left unjoined because immersion in the molten metal completes the circuit and a rapid reading is obtained.

Fig. 5.7 Complete thermocouple arrangement

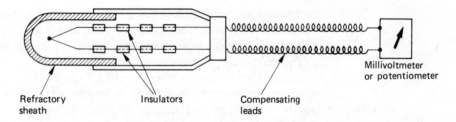

Refractory sheath   Insulators   Compensating leads   Millivoltmeter or potentiometer

Means must be provided for keeping the unconnected ends (the cold junction) of the thermocouple at a constant temperature. In the laboratory this can be conveniently achieved by immersing the wire-ends in a thermos-flask (remote from the heat source) containing pure melting ice. Thus the cold junction would be at 0°C. However, this arrangement is not convenient in industry, and the usual industrial practice is to transfer the cold junction by means of "compensating leads" to a position well away from the heat source, e.g. to a recording room. Compensating leads are wires, which have the same thermo-electric characteristics as the thermocouple wires. For chromel-alumel thermocouples copper-constantan compensating leads are used, which are connected to the e.m.f.-measuring instrument (fig. 5.7). The latter is set to the actual temperature of the cold junction (usually room temperature) if the instrument is calibrated directly in degrees Celsius.

Fig. 5.8 Potentiometer circuit for measuring thermocouple e.m.f.

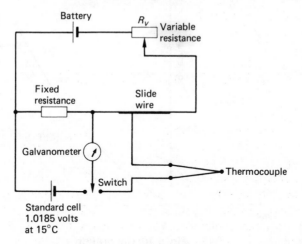

As indicated previously the e.m.f. generated by the thermocouple is measured by a millivoltmeter or a potentiometer. The millivoltmeter should have high electrical resistance and may be of the indicating type or it may be chosen to produce a permanent record. The more accurate method of measuring e.m.f. is by the use of a *potentiometer*, the principle of which is shown in fig. 5.8. The circuit is standardised against the standard cell by adjusting the variable resistance R_v until the galvonometer shows zero reading. The standard cell is then switched out and the thermocouple brought into the circuit. The e.m.f. generated by the couple deflects the galvanometer which is brought back to zero by adjusting the calibrated slide wire. Thus the thermocouple e.m.f. is determined and can be converted into a temperature by the use of tables. The standardisation and balancing of the e.m.f. in the circuit is achieved automatically in industrial potentiometers.

The thermocouple must be calibrated under similar circuit conditions (with compensating leads) as when it will be used. The couple can be checked against a standard thermocouple of known accuracy or against the freezing points of certain pure metals. This is carried out by melting the metal carefully (to prevent oxidation) in a small refractory crucible, and then transferring the crucible into a container which is well-lagged with glass wool to ensure slow cooling. The thermocouple with its sheath is inserted into the crucible of molten metal and e.m.f. readings taken at intervals of, say, 30 seconds as the metal cools. A graph is then plotted of millivolts against time to give a direct cooling curve for the particular metal (fig. 5.9). The horizontal portion of the curve gives the e.m.f. corresponding to the freezing point of the particular metal. Alternatively an inverse-rate cooling curve may be plotted by taking the time interval necessary for a decrease of unit e.m.f. reading. Resulting from this, a series of e.m.f. values corresponding to the freezing points of the pure metals used is obtained and a calibration graph for the thermocouple can be plotted (fig. 5.10).

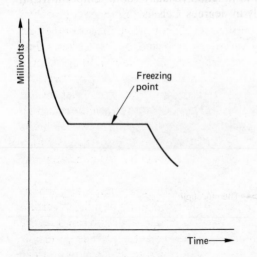

Fig. 5.9 Direct cooling curve of a pure metal

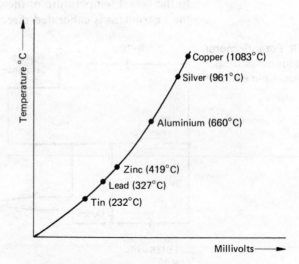

Fig. 5.10 Calibration curve for a thermocouple

Thermocouples may be subject to **errors** apart from false readings resulting from inaccurate calibration. These errors may arise from a number of causes including the following:

1 A common problem is that of contamination of the thermocouple wires which can result in embrittlement of the wire and alteration of the thermo-electric effects.

2 Bad connections or electrical leakages in the circuit.

3 Inadequate control of the cold junction temperature.

4 The positioning of the thermocouple is extremely important and there may be a significant time lag for the couple to show the correct temperature, especially where thick sheaths are used. Such a time lag is obviously a problem where the temperature fluctuates. The couple must extend to a reasonable distance inside a furnace and horizontally mounted couples must be properly supported. Further, couples should be positioned so that they are not directly in the path of a flame.

5 The measuring instrument must be sited so that it is not affected by vibration or heat. The compensating leads must be long enough to enable the instrument to be positioned well away from the furnace.

5.4 Non-contact Pyrometers

A **non-contact pyrometer** is used to measure high temperatures which are beyond the range of thermocouples or to measure the temperature of a surface on which it is inconvenient or impossible to place a couple. In such cases an instrument is used which measures the intensity of radiation emitted from the hot source.

When the temperature of an object is raised, the total amount of radiation increases and it is emitted at shorter wavelengths. This gives two visual effects: the object appears brighter and there is a change in the colour of the light emitted. The first effect is used for measurement, while the second effect enables temperatures to be very approximately judged by colour. For example, an object between 700°C and 800°C is usually described as "cherry red", between 800°C and 900°C as "orange", and at about 1200°C as "white". Radiation pyrometers are not concerned with the colour of the radiation but with its intensity and are of two main types: *total radiation* pyrometers, which make use of *all* the energy radiated, including the longer wavelengths ("infra-red") to which the eye is not sensitive, and *optical* pyrometers which deal only with visible radiation and rely on visual matching for measurement of brightness.

A third type of pyrometer, namely the *photoelectric* pyrometer, also makes use of the radiation emitted by hot objects. When radiant light falls on certain materials (e.g. silicon, selenium) electrons are released and an electric current flows in proportion to the intensity of the radiant light. The current is measured by a galvanometer which may be calibrated to give a temperature reading directly. This type of pyrometer is simple in construction and has a rapid speed of response to temperature changes, so that it is suitable for use in automatic temperature control equipment. The pyrometer may be used in the temperature range 800–3000°C.

Principles of Radiation Pyrometers

All materials when heated to the same temperature do not emit the same amount of radiation, and at a given temperature there is a maximum amount of energy which can be radiated from the surface of a given material. If a hot object radiates the maximum possible energy, in relation to its temperature, and also absorbs all radiation falling on it, then the object is termed a *black body* (see also section 5.8). The assumption is that no radiation is lost due to reflection from the body or transmission through the body. The black body represents an ideal state which is taken as the standard when dealing with the radiation characteristics of hot objects. Ordinary hot objects or sources such as furnaces are not ideal black bodies because they reflect some of the radiation and radiate less than the possible maximum. Black-body conditions are approached when a hot object is viewed through a small aperture, because little energy is lost, and it is good practice during furnace heating to take the temperature of components through a small hole in the furnace.

The fraction of the total energy radiated per unit area from an object at absolute temperature T as compared with an equivalent black-body surface at the same temperature is called the *emissivity* ϵ of the surface. If the hot object behaves as a black body its ϵ value will be 1. The value depends on the nature and colour of the surface as well as on its temperature. Matt (rough) surfaces approach black-body conditions more closely than do smooth surfaces regardless of colour; also oxides usually have higher ϵ values than metals, and, generally, the nearer to black the colour is, the closer the approach to black-body conditions. In industrial practice, hot objects absorb and reflect part of the radiation and emit only a small fraction of the radiation which would be emitted by a black body. As mentioned previously, when a metallic object is heated its colour changes as its temperature rises and the amount of heat radiated increases. Therefore, the emissivity of a metal varies with temperature and also with the wavelength of the radiation. It is clear therefore that emissivity corrections are necessary in practice in order to obtain the correct temperature.

Fig. 5.11 Féry total radiation pyrometer

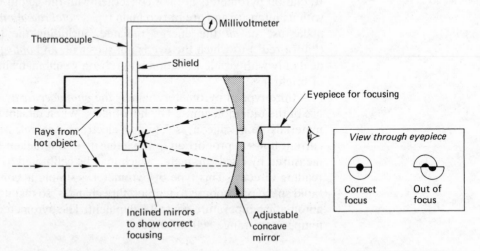

In the **total radiation pyrometer** type, a known fraction of the radiation of the hot body is collected and measured. The *Féry radiation pyrometer* (fig. 5.11) is a widely used instrument of this type. In this pyrometer, radiation from the hot object is accurately focused, by means of a concave mirror, on to a thermocouple mounted at the point of focus. The position of the concave mirror is adjusted until the image of the two inclined mirrors in front of the thermocouple is seen as a single circle. The focus is incorrect if two halves of a circle are observed out of line. A small polished shield protects the thermocouple from unfocused radiation which may cause erroneous readings. A black target is fixed to the hot junction so that it readily absorbs the heat focused on to it. The thermocouple is connected to a millivoltmeter which may be calibrated to give the temperature directly. The pyrometer requires about 15 seconds for a reading to be taken, but can be designed to give a permanent record. The readings are practically independent of the distance of the instrument from the hot body, and the size of the hot body, so long as the image of the hot body formed by the concave mirror in the instrument is large enough to cover the thermocouple.

As stated previously, total radiation pyrometers measure all wavelengths of the emitted radiation, and used under black-body conditions are subject to the *Stefan-Boltzmann Law*, which states that the total heat energy radiated by a body is proportional to the fourth power of its absolute temperature:

$$Q = \sigma(T^4 - T_0^4)$$

where Q = total heat energy radiated
T = absolute temperature of radiating surface
T_0 = absolute temperature of surroundings
σ = a constant.

In the case of a metallurgical process, T_0^4 is usually very small compared with T^4, so that the expression may be simplified to

$$Q = \sigma T^4$$

The observed temperature must be corrected for the emissivity of the hot object, and because the whole spectrum of emitted radiation is measured the total emissivity factor ϵ is used.

If T = true absolute temperature, then

$$Q = \sigma \epsilon T^4$$

and if S = observed absolute temperature, then

$$Q = \sigma S^4$$
$$\sigma \varepsilon T^4 = \sigma S^4$$
or $\qquad T = S/\epsilon^{\frac{1}{4}}$

Hence if the observed temperature of a steel surface were 1200°C, the true temperature, assuming the total emissivity of iron at 1200°C to be 0.32, would be given by:

$$T = \frac{(1200 + 273)}{(0.32)^{\frac{1}{4}}} = \frac{1473}{0.75} = 1958 \, \text{K} = 1685°C$$

A common form of **optical pyrometer** is the *disappearing filament pyrometer* (fig. 5.12). In this type the intensity of light radiation of a given wavelength emitted by the hot body is compared with light at the same wavelength from a standard source. The standard source brightness is obtained from a small incandescent lamp, the filament of which is placed at

Fig. 5.12 Optical pyrometer

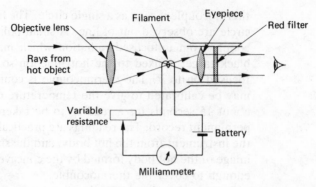

Objective lens   Filament   Eyepiece   Red filter

Rays from hot object

Variable resistance

Battery

Milliammeter

View through eyepiece

| Hot object brighter than filament, i.e. reading too low | Filament blends into background, i.e. reading correct | Filament brighter than hot object, i.e. reading too high |

the focal point of an objective lens. The current passing through the filament is adjusted by a variable resistance. The radiation from the hot body passes through the objective lens and a red glass filter only allows light of wavelength 0.65×10^{-6} m (0.65 micron) to pass into the eyepiece. When viewing the hot body through the eyepiece, the filament current is adjusted until the filament disappears against the background of the hot body. The radiation emitted by the latter is then the same as that of the filament. The current passing through the filament is measured on a milliammeter which is calibrated directly in temperature. The observed temperature must be corrected for the emissivity value of the hot body at a wavelength of 0.65 micron.

The principle on which the optical pyrometer is based is *Wien's Law*, which can be expressed as follows:

$$I = c\lambda^{-5}e^{-k/\lambda T}$$

where I = intensity of radiation emitted by hot body
       λ = wavelength of radiation (usually 0.65 micron)
       T = absolute temperature of hot body
  c, k = constants
       e = base of natural logarithms (2.718).

For non-black body conditions suppose

     S = observed temperature (K)
     T = true temperature (K)
     ϵ = emissivity at 0.65 micron and at temperature S.

then

$$c\epsilon\lambda^{-5}e^{-k/\lambda T} = c\lambda^{-5}e^{-k/\lambda S}$$
$$\epsilon e^{-k/\lambda T} = e^{-k/\lambda S}$$

Taking logs to base e

$$\log_e\epsilon - \frac{k}{\lambda T} = \frac{-k}{\lambda S}$$

$$\frac{k}{\lambda}\left(\frac{1}{T} - \frac{1}{S}\right) = \log_e\epsilon$$

or $\quad \dfrac{1}{T} - \dfrac{1}{S} = \dfrac{\lambda}{k}\log_e\epsilon$

from which the true temperature can be found.

The difference between the true and observed temperature is much less than in the case of the total radiation pyrometer, because the optical pyrometer measurement is carried out at one wavelength (0.65 micron).

The disappearing filament pyrometer is suitable for use in the temperature range 800–1300°C, above which the hot body and the filament are too bright for accurate matching. However, the range may be extended by the use of suitable filters.

The optical pyrometer is a portable instrument and it is not possible to have continuous readings from it. Therefore it cannot be used for automatic temperature control. Its greatest use is for spot checks on objects in furnaces, products during hot working and streams of molten metal during tapping.

Radiation pyrometers are suceptible to **errors** due to the emissivity of the hot object (e.g. solid metals may be coated with oxide films, while molten metals may be covered by slag). Any gases, vapours or smoke existing between the hot object and pyrometer may result in radiation being absorbed. False readings can also be obtained if radiation from hotter surfaces near to the object (whose temperature is being determined) is collected by the pyrometer. In optical pyrometers each reading is the result of a photometric match that is dependent on the judgement of the operator, thus introducing a possible human error. Further, the electric lamp in an optical pyrometer is susceptible to ageing which causes false readings.

5.5 Heat Transfer

Heat may be regarded as energy in transit between two bodies because of a difference in temperature. When two bodies are brought into contact, the vibrational energy of the atoms of the hotter body tend to decrease and that of the atoms of the cooler body to increase until both are at the same temperature. There is a transfer of energy from the hotter to the cooler body and energy transferred in this way is called heat. The term "heat" applies only to energy in transit and cannot be used to describe stored energy. The energy possessed by the atoms of a body is called **internal energy** so that a cooler body receiving heat from a hotter body stores the energy received as "internal energy". **Heat transfer** tends to occur wherever there is a temperature difference and there are three ways in which such energy may be transferred: conduction, convection and radiation.

The laws of heat transfer are of controlling importance in the design and operation of furnaces, heat exchangers, etc. In some equipment the main objective is to achieve the maximum heat transfer rate per unit area of surface, whereas in other cases it may be to recover heat from waste gases or to minimise heat losses by insulation.

5.6 Conduction

Conduction in a homogeneous opaque solid is the transfer of heat from one part to another under the influence of a temperature gradient, without any visible motion of the body. Conduction involves the transfer of energy from one atom to an adjacent atom. It is the only method of heat flow in an opaque solid body. (With some transparent solids, e.g. quartz, some energy is transmitted by radiation as well.)

Thus, if one end of a metal rod is held in the hand and the other end is heated, the hand can detect the flow of heat by conduction along the rod. Atoms with high vibrational energy at the hotter end of the rod transfer energy directly to lower energy atoms with which they are in contact, and these in turn transfer energy to adjacent atoms. This is repeated for layer after layer of atoms until the other end of the rod is reached. Each layer of atoms is at a slightly lower temperature than the preceding one, i.e. there is a temperature gradient along the rod. If the temperatures at either end of the rod remain constant, then the rate of heat transfer reaches a **steady state condition** at which the heat transfer rate does not vary with time or position in the system.

Experimental work shows that the amount of heat transferred per unit time, Q, by conduction through a flat plate of material of thickness x and area A, whose faces are kept at uniform temperatures T_1 and T_2, is inversely proportional to x, and directly proportional to A and to $(T_1 - T_2)$. Hence

$$Q \propto \frac{A(T_1 - T_2)}{x} \quad \text{or} \quad Q = \frac{kA(T_1 - T_2)}{x}$$

where k = a constant for the material, known as the **thermal conductivity** of the material. Therefore

k has the units $(J \times m)/(m^2 \times s \times °C)$ or $Wm^{-1}(°C)^{-1}$

Metals have high thermal conductivities (e.g. Ag 419, Cu 388, and Al 203 $Wm^{-1}°C^{-1}$), whereas non-metals usually have low values (e.g. cork 0.04 $Wm^{-1}°C^{-1}$) and their poor conductivities may permit their use as heat insulators (i.e. poor conductors).

Re-arranging the above equation

$$Q = kA\frac{(T_1 - T_2)}{x}$$

where Q = rate of heat transfer

$(T_1 - T_2)/x$ = temperature gradient.

Example   Calculate the amount of heat lost per sec by conduction through a copper tank wall, $2\,m^2$ in area. The temperature difference between the inside and outside of the wall is 20°C. The copper is 1 cm thick and the thermal conductivity of copper is $380\,Wm^{-1}°C^{-1}$.

$$Q = \frac{kA(T_1 - T_2)}{x} = \frac{380 \times 2 \times 20}{0.01} = 1.52 \times 10^6\,J$$

Conduction through a Composite Wall

In many practical cases heat transfer involves more than one material. For example, heat may escape from a furnace by travelling first through a layer of refractory brick and then through a layer of insulating brick of different thermal conductivity. The calculation of the rate of heat transfer in such cases depends on the fact that the total amount of energy transferred during a given period is the same for all layers. For example, in the case of a composite wall consisting of three layers of different materials as shown in fig. 5.13.

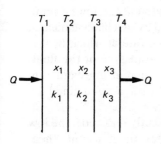

Fig. 5.13

$$Q = \frac{k_1 A(T_1 - T_2)}{x_1} = \frac{k_2 A(T_2 - T_3)}{x_2} = \frac{k_3 A(T_3 - T_4)}{x_3}$$

$$T_1 - T_2 = Qx_1/k_1 A$$
$$T_2 - T_3 = Qx_2/k_2 A$$
$$T_3 - T_4 = Qx_3/k_3 A$$

$$\therefore\quad T_1 - T_4 = \frac{Q}{A}\left(\frac{x_1}{k_1} + \frac{x_2}{k_2} + \frac{x_3}{k_3}\right)$$

Example   A furnace wall consists of firebrick, insulation brick and building brick, the thickness of the latter being 0.25 m. The temperature on the furnace side of the firebrick is 1000°C. The maximum temperature to which the insulation brick is subjected is 800°C and the outside temperature of the building brick is 25°C. Calculate the thickness of the firebrick and insulation brick if the rate of heat transfer through the composite furnace is $400\,W/m^2$.

Thermal conductivity of firebrick $= 1.25\,Wm^{-1}°C^{-1}$

insulation brick = 0.25

building brick   = 0.70

Consider $1\,m^2$ of wall, then

$$400 = 1.25(1000 - 800)/x_1$$
$$x_1 = 1.25 \times 200/400 = 0.625\,m$$
$$\therefore \text{ Thickness of } firebrick = 0.625\,m$$

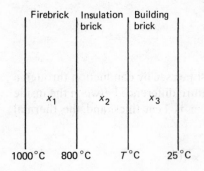

Firebrick | Insulation brick | Building brick

x_1 x_2 x_3

1000°C   800°C T°C   25°C

Fig. 5.14

Considering the building brick:

$$400 = 0.7(T - 25)/0.25$$
$$100 = 0.7(T - 25)$$
$$143 = T - 25$$
$$\therefore T = 168°C$$

Considering the insulation brick:

$$400 = 0.25(800 - 168)/x_2$$
$$400x_2 = 0.25 \times 632$$
$$x_2 = 158/400 = 0.395 \text{ m}$$
$$\therefore \text{ Thickness of } insulation \text{ } brick = 0.395 \text{ m}$$

5.7 Convection

The term **convection** is used when heat is transferred by the movement of matter. Thus, if one region of either a gas or a liquid is heated it will become less dense than the surrounding gas or liquid. This will result in the heated substances being displaced upwards. Heat is thus transferred to other regions by movement of the fluid. This is *natural convection*. Water in some small heating installations circulates by convection. A boiler heats the water, which then rises through a pipe into radiators. As the water cools so it falls back through a return pipe to the boiler which is situated at the lowest point of the system. The motion of the fluid may be produced by mechanical means, e.g. by fans and pumps. This is *forced convection*. The rate at which heat is transferred by forced convection depends on the speed of flow of the fluid. Thus, if a fan blows air over a heated object, it loses heat at a greater rate than if the air were still. Convection is almost always accompanied by conduction.

If natural convection can be prevented (or at least greatly reduced), the low thermal conductivity of air, and other fluids, makes them useful "heat insulators". In such insulating materials as wool, glass fibre and foamed plastics, the function of the material is to separate the air which it contains into small pockets so that normal convection cannot take place; it is this air which is, in fact, the insulator.

5.8 Radiation

Conduction and convection both depend on direct contact, either between two bodies or between a body and a fluid. **Radiation** differs from these modes in that no contact of any kind is required; the energy is transmitted in the form of electromagnetic waves which can travel through a vacuum. Heat radiation is emitted by all bodies which are above the absolute zero of temperature. The rate at which energy is emitted depends on the temperature of the body, its surface area, and the nature of the surface. The temperature of the body also determines the range of wavelengths of the radiation and at high temperatures the body emits visible light.

Experiment shows that there is a relationship between the ability of a surface to *emit* radiation and its ability to *absorb* it. Good absorbers are also good emitters. Thus a polished metal surface reflects most of the radiation falling on it and absorbs very little, so that it is also a poor radiator. On the other hand, a surface which appears black does so because it absorbs most of the light falling on it, and it would also absorb most of the radiation. A body which absorbs all the radiation falling on it is called a *black body*, and it follows that such a body is an ideal radiator.

Experimental work also shows that, for a black body, the total energy E emitted per unit time is proportional to the area A of the body and to the fourth power of its absolute temperature T.

$$E \alpha A T^4$$
$$\text{or} \quad E = \sigma A T^4$$

This is the Stefan-Boltzmann law, and σ, the Stefan-Boltzmann constant, is found to have the value $5.67 \times 10^{-8} \, \text{Wm}^{-2}\text{K}^{-4}$.

If the body is not "black", less energy will be radiated and the equation becomes

$$E = \epsilon \sigma A T^4$$

where ϵ is called the emissivity of the surface. The value of ϵ varies from nearly 1 for lamp black to about 0.02 for polished silver; clean copper has an emissivity of 0.1, while the value for oxidised copper is about 0.7.

If two bodies, one hotter than the other, are placed in an enclosure then there will be a continuous intercharge of energy between them. The hotter body radiates more energy than it absorbs, while the colder body absorbs more than it radiates. Even when equilibrium and equalisation of temperatures has taken place the process continues, each body radiating and absorbing energy. The net heat transfer for any particular body is the difference between the energy radiated and the energy absorbed. For example, when a furnace and its charge have reached a uniform temperature each will be radiating energy but the net heat transfer will be zero. Thus, if a black body of area A at absolute temperature T_1 is completely surrounded by a black surface at absolute temperature T_2, then in unit time it will radiate energy $\sigma A T_1^4$ and absorb energy $\sigma A T_2^4$, so that the rate of heat transfer will be

$$Q = \sigma A (T_1^4 - T_2^4)$$

Example Calculate the energy loss per second by radiation from a surface of area $0.2 \, \text{m}^2$ when it is at a temperature of 727°C and the surroundings are at 27°C ($\sigma = 5.67 \times 10^{-8} \text{Wm}^{-2}\text{K}^{-4}$).

Total energy lost by the surface per second

$$= 5.67 \times 10^{-8} \times 0.2 \times (1000^4 - 300^4) \, \text{J}$$
$$= 5.67 \times 10^{-8} \times 0.2 \, (10^{12} - 3^4 \times 10^8) \, \text{J}$$
$$= 5.67 \times 0.2 \, (10^4 - 81) \, \text{J}$$
$$= 1.13 \times 10^4 \, \text{J}$$

Heat transfer in practice can rarely be identified as being solely due to one mode, but tends to occur by a combination of modes. Fig. 5.15 indicates the modes of heat transfer in a fused salt bath.

Fig. 5.15 Heat transfer in a fused salt bath

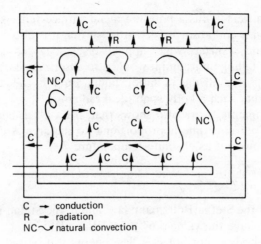

C → conduction
R → radiation
NC ⌇ natural convection

5.9 Recovery of Waste Heat

In processes using combustible fuels the heat from the emerging waste gases may be recovered, thus improving the thermal efficiency of the process. The heat recovered may be used to preheat both the fuel and the air required for combustion, thus resulting in fuel saving and/or higher processing temperatures. Heat recovery methods include regeneration, recuperation and the use of waste heat boilers.

Regeneration is an intermittent process in which the hot waste gases are passed through large chambers filled with a honeycomb of firebricks (called checker brickwork) to which they give up much of their heat. After about thirty minutes the hot waste gases are diverted to another regenerator chamber which has been preheating the air for combustion; the latter is now passed through the newly heated regenerator and the regular alternation is continued. If both air and gaseous fuel are being preheated then two sets of double-compartment regenerators are required. The principle of regeneration is used to preheat the air supplied to blast furnaces and the regenerators are called hot blast stoves (see section 2.5).

Recuperation is based on conduction through solid material which separates the fluid to be heated from the fluid which is cooled as illustrated in fig. 5.16. Severe demands are made on the recuperator tube material which is in contact with very hot gases. The tube wall must be thin and of high thermal conductivity. However, after being in use for some time metallic tubes may become oxidised on their inner surfaces resulting in a lowering of their ability to transfer heat. In order to ensure satisfactory life, tubes are made of

Fig. 5.16 Recuperation methods

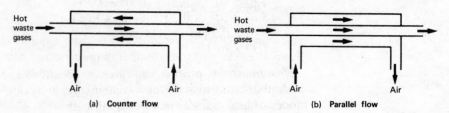

(a) **Counter flow**   (b) **Parallel flow**

expensive heat-resisting materials and both metallic and refractory materials are used.

Heat recovery from waste gases by regeneration and recuperation is only partially effective and further heat may be recovered by using waste heat boilers to generate steam. This steam may be used for space heating or power production. Heat absorption by the surfaces occurs mainly as a result of forced convection so that adequate draught is required to ensure turbulent flow. The design of boiler specified for a particular purpose will depend on:

a) The temperature and volume of waste gas available.
b) The nature and concentration of dust content.
c) The corrosive nature of the gas.
d) The demand for process steam, space heating and power.

5.10 Insulating Materials

All materials conduct heat to some extent, so the term "insulator" is used to describe a relatively poor conductor. The main requirements of an **insulating material** are that it must have:

1   Low thermal conductivity.
2   Good resistance to heat, including high charring temperature and high ignition temperature.
3   Good formability so that it can be easily applied.
4   Good durability to ensure long life.
5   Good resistance to attack (e.g. rot-proof and vermin proof).
6   Low cost.

A material which is a poor conductor of heat owes its properties to the porous nature of its structure. For example, an insulating refractory brick contains up to about 70% porosity. In order to be most effective these voids must be totally enclosed and so small that convection within them, and radiation across them, is minimised. The transfer of heat by conduction will be very small because of the low thermal conductivity of air, and radiation becomes less significant as the temperature of the void surfaces falls. Therefore the overall rate of heat transfer through such a material will be low.

The raw materials used in producing insulating materials include the following.

White asbestos occurs as a rock and is a fibrous material. The fibres are easily separated into various lengths. It is used alone or in conjunction with other materials and can be pressed into various shapes (e.g. blocks, board).

Magnesium carbonate is a light, white powder which is used in conjunction with about 10% white asbestos for lagging steam pipes, and is suitable for temperatures up to about 300°C.

Vermiculite, which is a micaceous mineral, is used in the form of a plaster, made up of 60% vermiculite, 30% calcium sulphate and 10% white asbestos; this plaster can withstand temperatures up to red heat without disintegrating.

Diatomaceous earth (diatomite or kieselguhr) is a naturally occuring hydrated silica which is granular and very porous (about 90% of the volume). This material can withstand temperatures up to about 850°C.

Mineral wool (slag wool or rock wool) is made by disintegrating blast furnace slag with high-pressure air or steam jets. The high-pressure jet breaks the molten slag stream into small globules and propels them at high speed so that fibres are formed. The wool is then compacted into shapes such as blankets, blocks, pipe covering.

The **advantages of insulation** are:

1   Saving of fuel, with side benefits such as improvement in product quality due to more accurate temperature control and more uniform temperature distribution.

2   Protection of refractories from rapid changes in temperature, thus helping to prevent spalling of the brickwork.

3   Improvement of the working conditions around the furnace due to the confinement of heat within the furnace.

4   The insulation of steam pipes helps to prevent condensation and allows delivery of drier steam as well as minimising corrosion. The insulation of bare metal surfaces is even more important than furnace insulation due to the high thermal conductivity of metal.

The **disadvantages** of insulation are:

1   The hot-face temperature of refractory bricks will be higher, so that the surface will be softer, resulting in greater wear and increased rate of attack.

2   The insulating material may be bulky, resulting in larger furnace dimensions.

3   Insulation is costly and is only justified when it gives a cost-saving by reducing fuel consumption.

Exercises 5

1   Explain what is meant by the "Kelvin temperature scale".

2   *a*) Name the three essential parts of a complete thermo-electric pyrometer.
b) Name a base-metal type of thermocouple capable of continuously measuring a temperature of 1100°C.
c) State two advantages of this thermocopule over a noble metal thermocouple.
d) State three precautions which may be taken to avoid errors in the use of thermocouples.

3   *a*) Name two types of radiation pyrometer.
b) With the aid of a sketch explain how one of the above pyrometers works.
c) Name two errors to which radiation pyrometers may be subject.

4   *a*) Explain what is meant by the following terms:
(i) black body, (ii) emissivity
b) The observed temperature of a metal surface is 1300°C. Assuming the total emissivity of the surface to be 0.33, calculate the true temperature.   [*Ans*. 1800°C]

5   *a*) Explain what is meant by "steady state" heat transfer in conduction.
b) A furnace wall consists of a layer of refractory brick 100 mm thick and an outer layer of building brick 200 mm thick. The inner and outer surfaces of the composite wall are at 600°C and 50°C respectively, and the thermal conductivities of the refractory and building brick are respectively 1.4 and 0.6W/m°C. Calculate *a*) the temperature at the junction of the two materials, *b*) the rate of heat transfer per m^2 of the composite wall.   [*Ans*. *a*) 485°C, *b*) 1.61 kW]

6   *a*) State the Stefan-Boltzmann law regarding the radiation from a hot object.
b) A cylindrical steel forging 0.3 m diameter and 2 m long is supported at its base while being cooled down during a normalizing heat treatment. At what rate is it losing heat by radiation when its surface temperature is 600°C? The temperature of its surroundings is 20°C and it may be assumed to be a "black body". The Stefan-Boltzmann constant is 5.67×10^{-8} W/m^{2}K^4.   [*Ans*. 70.2 kW]

7   *a*) State two advantages of heat insulation of a furnace.
b) Name two desirable characteristics of an insulating material.
c) Name two insulating materials suitable for use at 1000°C.

6 Refractory Materials

Refractories are essentially non-metallic materials which can withstand the high temperatures and arduous conditions encountered in metallurgical operations. They must have the ability to

a) maintain their dimensions, strength and rigidity at the operating temperatures;

b) withstand the thermal shock associated with rapid change of temperature;

c) resist chemical attack by gases, liquids (molten metal and slag) and solids with which they come into contact.

There is no one refractory material which is able to withstand all possible conditions, so that a choice must be made to meet the main requirements of the particular application. In a furnace, the brickwork must confine the heat within the furnace in order to conserve fuel and protect the outer steel shell of the furnace. The materials used for crucibles, retorts and muffles should have relatively high thermal conductivity, whereas insulation refractories must have low thermal conductivity. In waste-heat recovery plant, heat may be stored in regenerators, which thus need to be constructed of material of good thermal capacity; waste heat may also be recovered by recuperators made of material with high thermal conductivity.

Because of the method of manufacture a refractory material will contain pores and the porosity may vary from 10 to 70%—a typical value for silica brick being about 25%. Open pores make the material permeable and allow attack by the penetration of liquids and gases; closed pores (which are sealed off during the firing stage of manufacture) are required if the refractory is intended for insulation purposes. Therefore, the control of porosity (and hence permeability) is important. Compressive stresses may be imposed by the surrounding furnace structure and by the charge, so that the hot strength of the material is an important property. The impingement of rapidly moving solid particles of charge and slag held in suspension in the furnace gases results in the refractory being subjected to an erosive action which must be resisted.

It follows from the above that in selecting a refractory material for a particular purpose due attention must be paid to the chemical, physical and mechanical properties.

6.1 Classification of Refractories

Refractories may be classified according to their chemical reactivity into acid, basic and neutral types.

Acid refractories have silica (SiO_2) as their main constituent, but alumina

146

(Al_2O_3) may also be present in substantial amounts. Highly siliceous materials containing alumina may be subdivided recording to their content of Al_2O_3 as follows:

a) Siliceous bricks; Al_2O_3 content less than 25%
b) Firebricks; Al_2O_3 content 25 to 40%
c) Aluminous bricks; Al_2O_3 content more than 40%.

Basic refractories are based on MgO and include magnesite, dolomite, and chrome-magnesite. In addition, alumina and mullite ($3Al_2O_3.2SiO_2$) are classed as basic materials. MgO is one of the most refractory of the common oxides, having a melting point of 2800°C, compared with about 1700°C for SiO_2 and 2050°C for Al_2O_3. The melting point of CaO (2570°C) approaches that of MgO, but CaO reacts with moisture to produce $Ca(OH)_2$ and crumbles in the process.

The MgO is produced by calcining magnesite rock:

$$MgCO_3 \rightarrow MgO + CO_2$$

Dolomite rock ($CaCO_3.MgCO_3$) is treated similarly to produce CaO.MgO. In the latter material the CaO is stabilised by adding a small amount of siliceous material before the firing process; the CaO does not then react with moisture.

Neutral refractories do not have pronounced acidic or basic properties (e.g. graphite), or, alternatively, the acidic and basic properties are about equally balanced (e.g. chromite $FeO.Cr_2O_3$ and forsterite $2MgO.SiO_2$).

Acid refractories react readily with basic slags and basic refractories are usually attacked by acid slags. However neutral refractories are relatively inert to both siliceous and lime-containing slags. In furnace construction acid refractories should not be put next to basic refractories.

Super refractories are materials which have been specially developed for use at exceptionally high temperature (above about 1800°C) or for contact with very reactive material (e.g. in gas turbines). This group includes zirconia (ZrO_2), thoria (ThO_2), beryllia (BeO), silicon carbide (SiC), and various other carbides, nitrides and borides. Many of these materials are extremly brittle.

It should be noted that the general rules of using acid refractories with acid slags and basic refractories with basic slags are often broken in industry. For this reason refractories may be classified by means of the raw materials from which they are derived.

6.2 Testing of Refractories

The **testing of refractories** is carried out to control the quality of the material being made and also to assess the suitability of a given refractory for a particular use. Standard methods of testing refractories are covered by B.S. 1902.

The property of *refractoriness* is evaluated by assessing the softening temperature of the material. It is not usually practicable to measure the melting point of a refractory material because melting may extend over a relatively wide range of temperature. The test for refractoriness compares the sagging of a cone of the material (cut from a brick) with that of standard Seger cones when they are heated together at a standard rate (10–15°C per minute) in a mildly oxidising atmosphere until the test cone bends over (fig. 5.1, page 125). The number of the best matching Seger cone is quoted as the refractoriness of the brick. The maximum temperature at which the refractory should be used is well below the refractoriness figure. Also, in practice, reducing gases may be present in the furnace and may cause a significant decrease in refractoriness.

In furnace construction the brickwork is subjected to stress arising from the weight of the surrounding brickwork, or from the thrust of an arch, as in a roof, or pinching due to expansion. These stresses may cause the bricks to deform at a temperature much below their normal refractoriness. The determination of the **refractoriness-under-load** involves the assessment of high temperature deformation of the refractory when subjected to a specified compressive load. Two forms of the test are used:

1 The *rising temperature test* which involves heating the loaded test-piece at a prescribed rate until either collapse or a specified amount of deformation occurs.
2 The *maintained temperature test* in which the loaded test-piece is heated at a prescribed rate to a pre-determined temperature which is maintained either for a specified time or until collapse or a pre-determined deformation occurs.

Test 2 is considered to provide information which is more relevant to industrial conditions than test 1. However, refractoriness-under-load testing is tending to be replaced by the modulus of rupture test which is considered to give equally relevant results with far less effort.

The **modulus of rupture test** is carried out on heated refractories, both as an index of brick quality and as an indication of probable performance in service. A compression testing machine capable of exerting a load up to about 100 kN is suitable and a uniform loading rate of about 50 kN/m^2 per sec is used. The test piece may be dried whole bricks 150 × 25 mm or specimens cut to this size from a larger shape. The specimen is supported on bearing edges and loaded at the centre. The distance between the bearing edges is usually about 125 mm but is not critical. The temperature is measured by a platinum-platinum/rhodium thermocouple. The test-piece is tested to destruction and the modulus of rupture calculated from the formula:

$$\text{Modulus of rupture} = \frac{3Wl}{2bd^2} \text{ kN/mm}^2$$

where W = load at failure (kN)
l = distance between bearing edges (mm)
b = breadth of specimen (mm)
d = depth of specimen (mm).

Modulus of rupture figures for refractory materials are quoted over a wide range of temperatures.

The **resistance to thermal shock** measures the ability of a refractory material to withstand the destructive effects of sudden changes in temperature. There is no simulative test which could serve as a standard method, but materials can be compared by using constant test conditions which should be chosen to suit the brick under test and the anticipated working conditions. A *spalling* test is often used—spalling being defined as the breaking or cracking of the brick, often at the surface, so that pieces fall away leaving fresh surfaces exposed.

A useful general-purpose spalling test uses a muffle furnace of such a size that when the cooled test pieces are inserted the fall in temperature does not exceed 20°C and the test temperature is regained within 5 minutes. Three test-pieces are cut in the shape of right prisms 75 mm high with a square base of 50 mm side. The test-pieces are thoroughly dried before being tested. They are then placed in the cold furnace which is heated at a uniform rate so that it attains the test temperature in 3 hours. For silica bricks a temperature of 450°C is used and for other refractory bricks 1000°C is chosen. The testing temperature is maintained for 30 mins and the test-pieces are then removed from the furnace with warm tongs and placed on a steel plate at room temperature for 10 minutes; after which they are returned to the furnace for 10 minutes and the above procedure repeated. Towards the end of each cooling period, the test-pieces are examined for loss of corners or detached pieces. The number of cycles necessary to cause detachment of pieces of brick is noted.

Resistance to slag attack is difficult to assess other than qualitatively. Numerous methods have been tried and reasonable correlations have been obtained with particular service conditions, but very few have been standardised or accepted for general use. One test involves drilling holes in the brick and packing these with samples of typical slag likely to be met. The brick is then heated to the working temperature for 1 hour, cut through and examined. The extent of the slag penetration into the brick is noted and compared with others. A slag test as above reveals trends in materials, but these must be subsequently confirmed by service trials.

Acid slags usually react vigorously with basic bricks and basic slags with acid bricks, but in practice prevailing condition are often more complex. For example, iron oxide, one of the more corrosive reagents, may occur in acid and basic slags and can damage both acid and basic bricks or be resisted by either type of brick in suitable circumstances.

Gases present in a furnace atmosphere may have a corrosive effect on the brickwork. For example, carbon monoxide at about 500°C in contact with firebrick may decompose, resulting in the deposition of carbon: $2CO \rightarrow C + CO_2$. The expansion caused by this deposition of carbon within the pores of the brick may be sufficient to shatter the brick.

The **density of a refractory brick** decreases as the firing of the material (during manufacture) proceeds to completion and is thus a guide to the extent of firing; the value is used for quality control purposes. For example, experience shows that silica bricks with a density of less than $2.4\,g/cm^3$ are satisfactory.

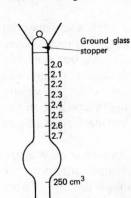

Fig. 6.1 Rees-Hugill flask

The relative density of powdered material may be determined by using a displacement flask (fig. 6.1) of the type designed by Rees and Hugill (covered by B.S. 2701). Accurate calibration of the flask is essential. The method is rapid in operation and is used for control testing, especially of silica refractories. The flask is filled to the lowest mark on the neck with xylene and allowed to stand for a few minutes to allow any liquid adhering to the neck to drain; the level of the xylene is finally adjusted by means of a narrow pipette inserted down the neck of the flask. Then, $100\,g \pm 0.1\,g$ of the test material is introduced into the flask, a little at a time, through a small funnel. When the whole quantity has been introduced, material adhering to the neck of the flask is dislodged by gentle tapping. The flask is then agitated without being allowed to become warm by contact with the hands, and after 30 seconds the relative density of the sample is read directly from the graduated neck of the flask.

As indicated earlier a refractory may contain open and/or closed pores. The **open or apparent porosity** is determined by measuring the volume of liquid which a known bulk volume of specimen will absorb. (Bulk volume is the total volume, including the total pore volume.) This is usually done by weighing the dried test-piece in air, W_a, soaking it and weighing suspended in the liquid, W_b, finally weighing the soaked test-piece while suspending in air, W_c. Then if D is the density of the liquid:

$$\text{Bulk volume } Y_a = \frac{W_c - W_b}{D}$$

$$\text{Bulk density } S_a = \frac{W_a D}{W_c - W_b}$$

$$\text{Apparent porosity } P_a = \frac{W_c - W_a}{W_a - W_b} \times 100(\%)$$

Penetration of liquid into open pores must be ensured and this may be achieved by evacuating the test-piece in a beaker in a vacuum desiccator and then adding the liquid to immerse the specimen whilst retaining the vacuum. Alternatively the test-piece may be heated to 110°C and plunged into a tank of boiling water—the steam evolved will flush the air from the pores.

The *true porosity P_t* is given by:

$$P_t = \left(1 - \frac{S_a}{S_t}\right) \times 100$$

where S_t is the powder density (found by using an S.G. bottle; or volume displaced by a known weight).

The volume percentage of closed pores is then the difference between P_t and P_a.

Usually some of the open pores are interconnected so that passages are provided through which gases or even liquids may pass. This constitutes **permeability**. A refractory's permeability is determined by measuring the volume of air which passes through a specimen of known dimensions in a given time with a known pressure drop across the specimen.

The permeability is given by the formula

$$P = \frac{VH}{tAp}$$

where V = volume of air (cm^3) passing in t seconds
H = height of specimen (cm)
A = cross-sectional area (cm^2)
p = pressure drop (cm of water).

Usually low porosity and permeability are required since

a) penetration of slags and fluxes lowers resistance to slag attack.
b) low permeability minimises leakage of gases from furnaces, gas passages, etc.

However it should be noted that the thermal conductivity of a refractory is dependent on the porosity—insulating (low thermal conductivity) bricks require high porosity.

Permanent linear change on reheating, often referred to as after-contraction or after-expansion, is of direct importance particularly with basic bricks since it may reveal soft-firing of the brick, or the presence of an undue amount of flux. In the case of dolomite bricks it provides an indirect measure of probable stability in storage—hard firing being the best guarantee of subsequent freedom from hydration troubles.

Samples cut from the brick are heated in a prescribed manner and, following cooling, are measured so that changes in dimension are reported as a percentage of the original length.

Reversible thermal expansion is of direct importance in that adequate expansion allowance should be made in furnace structure, but it is equally important in that it shows up peculiarities in expansion.

It is usually measured by comparing the linear expansion of a test piece approximately 10 cm long by 2 cm in diameter with that of fused silica (which has a low coefficient of expansion $0.54 \times 10^{-6}/°C$). One way of doing this is by means of an apparatus in which the test-piece rests on the bottom of a vertical closed-end silica tube, to the open end of which is clamped a dial gauge. On top of the test-piece rests a fused silica rod which actuates the dial gauge so that the expansion of the test-piece relative to the silica tube is registered on the gauge. The test-piece is heated at 50°C/min by inserting the lower end of the tube into a wire-wound furnace and the temperature is measured by a thermocouple placed alongside the specimen.

This method is unsuitable for use above 1200°C owing to the tendency of the fused silica to devitrify.

In an alternative apparatus, which permits determination to be made up to 1450°C or higher, a horizontal platinum-wound furnace is used and the expansion is measured by means of travelling telescopes fitted with vernier scales, which are focused on the ends.

Thermal conductivity is determined by subjecting a test-piece of known cross-sectional area to a given rate of heat flow and measuring the resulting temperature gradient. The thermal conductivity k is then found from the relationship:

$$k = \frac{\text{Rate of heat flow}}{\text{Area} \times \text{Temperature gradient}}$$

In measuring the thermal conductivity of a refractory material (with its relatively good insulating properties) the problem is to cause a measurable quantity of heat to flow through the test-piece. In the Lee's disc apparatus (fig. 6.2) this is achieved by applying heat to a test-piece of large cross-section and small thickness. Electrical power is supplied to the upper surface of the

Fig. 6.2 Lee's disc apparatus for determining thermal conductivity

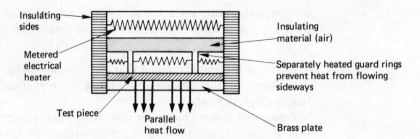

test-piece and the apparatus allowed to reach a steady state. The temperatures of the brass plates are effectively those of the test piece surfaces since temperature gradients in the brass are very small. The brass plate containing the electric heater is bounded by the test-piece on one side and on every other side by surfaces at the same temperature as itself. Therefore all the heat from the heater passes through the test-piece.

$$\text{Rate of heat flow} = \text{Volts} \times \text{Amps} = \frac{k \times \text{Area} \times (T_1 - T_2)}{\text{Thickness}}$$

where T_1 and T_2 are the observed temperatures (measured by thermocouples) of the test-piece surfaces.

$$\therefore \quad k = \frac{\text{Volts} \times \text{Amps} \times \text{Thickness}}{\text{Area} \times (T_1 - T_2)}$$

6.3 Composition, Properties and Uses of Common Refractories

Silica

The main raw material is quartzite, which contains 96–98% SiO_2 and should contain less than 0.3% of alkalies for brick making.

Silica brick has very good refractoriness under load at high temperatures and maintains its strength up to nearly fusion temperature. It has high resistance to thermal shock above 600°C, but very careful heating and cooling are necessary below 600°C because of the dimensional changes which occur within the polymorphic forms of silica, which exists in three principal crystalline forms. The low-temperature form is quartz, which is stable up to 870°C; between 870°C and 1470°C the stable form is tridymite; while from

1470°C to the melting point (1713°C), the stable form is cristobalite. The transformation from one crystalline form to another is known as the "inversion" of silica.

When quartz is heated to a temperature over 870°C for a sufficiently long time, it changes to tridymite or cristobalite according to the temperature. If the temperature is maintained between 870°C and 1470°C, any cristobalite which may have formed slowly disappears and tridymite is formed. The speed with which inversion occurs is greatly influenced by the presence of impurities, which accelerate the change; it is also affected by the particle size, a small size helping inversion. These inversions are normally very sluggish, so that it is possible for quartz, tridymite and cristobalite to exist side by side in a silica brick almost indefinitely at room temperature.

The inversion forms of silica have different relative densities as follows: quartz 2.653, tridymite 2.323, and cristobalite 2.318. Therefore, the inversion from quartz to tridymite or cristobalite is accompanied by considerable increase in volume; this expansion is usually referred to as permanent expansion. Also if the inversions are not complete in manufacture, the volume change is free to continue in service with the possibility of trouble resulting.

In addition to the major crystalline transformations, the three crystalline forms of silica themselves undergo reversible changes at definite temperatures. For example, quartz at room temperature (i.e. α quartz) changes to β quartz at about 573°C°; tridymite undergoes an inversion at about 117°C; while the cristobalite inversion occurs within the range 220–250°C. These reversible inversions are very rapid and are accompanied by distinct volume changes as shown in fig. 6.3. Since tridymite shows smaller sudden change

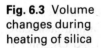

Fig. 6.3 Volume changes during heating of silica

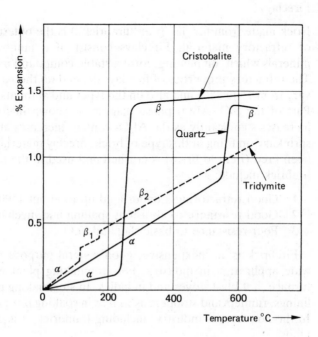

than cristobalite, it is desirable that during the firing of a silica brick as much tridymite as possible should be produced and bricks are available with over 90% of their crystalline silica existing as tridymite.

These reversible inversions, especially that of cristobalite, are of considerable importance in determining the resistance of a silica brick to sudden temperature changes. For example, a silica brick containing a large proportion of cristobalite, if quickly heated or cooled in the temperature range 200 to 300°C, will undergo sudden volume changes which will cause fragments of the brick to flake off. This phenomenon is known as "spalling".

The sensitivity of silica to abrupt changes of temperature restricts its use as a refractory material to those applications where sudden temperature fluctuations are absent. Silica bricks are used in roofs of reverberatory-type smelting and refining furnaces (e.g. in the extraction of copper). Another application is in the construction of the ovens used for the production of metallurgical coke.

Siliceous refractories are considerably cheaper than basic refractories.

Ganister

This is a sandstone in which the grains of SiO_2 are cemented together by clayey matter which, on grinding and moistening with water, is self-bonding. On firing, the volume change is small because the expansion of SiO_2 is offset by the contraction of the clay. Therefore it can be rammed and fired in position, and the absence of joints makes it a very suitable material for lining iron foundry cupolas and acid-lined converters.

Ganister is an inexpensive refractory material

Fireclays

Brick made from fireclay (i.e. "firebrick") is the oldest and commonest type of refractory material. Fireclays consist of a number of alumina-silicate minerals which, on heating, form a stable compound mullite ($3Al_2O_3.2SiO_2$). The refractory properties of fireclays depend on the Al_2O_3 content which can vary from 15 to 45% and also on the types and amounts of impurities present. Part of the SiO_2-Al_2O_3 phase diagram is shown in fig. 6.4. Refractoriness increases significantly as the Al_2O_3 content increases above 25%. To combat shrinkage on firing in this type of brick, fireclay material which has previously been fired (i.e. old brick) is crushed and used in the mix. The properties of firebrick include:

1 Good refractoriness under load up to about 1500°C.
2 Good resistance to abrasion, spalling and mechanical damage.
3 Poor resistance to basic slag and FeO.

Firebrick is an inexpensive, good general-purpose refractory which finds wide application in industry. For ironmaking plant it is used in the blast furnace, hot blast stoves and in ladles. In steelmaking plant it is used in ladle linings, runners and stoppers as well as in soaking pits and reheating furnaces. In the non-ferrous industry, including foundries, it is used in furnaces of all kinds.

Fig. 6.4 Part of SiO_2-Al_2O_3 phase diagram

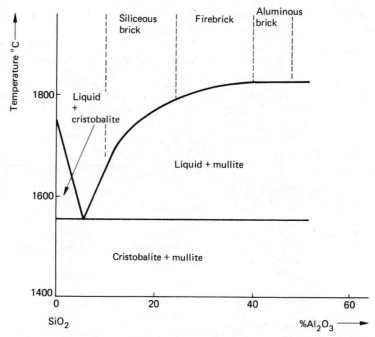

The quality of firebrick chosen for any given application depends on the prevailing conditions and the cheapest which will perform satisfactorily will obviously be used.

Magnesite

Magnesite rock consists of $MgCO_3$. U.K. deposits are of poor quality, and it is necessary to import most of the required supply (e.g. from Austria). This magnesite is calcined to give the very refractory oxide MgO:

$$MgCO_3 \rightarrow MgO + CO_2$$

In the manufacture of bricks a binding agent (e.g. a mixture of clay and tar) is added to the calcined magnesite before firing.

The *properties* of magnesite brick include:

1 Very good resistance to slags rich in CaO and FeO.
2 Good thermal conductivity.
3 Poor resistance to spalling (probably due to its high coefficient of thermal expansion).
4 Poor resistance to mechanical damage so that careful handling is necessary.

Magnesite bricks are expensive but are widely used in smelting and refining furnaces where basic slags and high temperatures are involved, e.g. steelmaking furnaces, reverberatory furnaces, and converters in non-ferrous smelting. Magnesite is invariably used in critical positions (e.g. hearths, tap-holes, top-runs of checker-work) in furnaces of all kinds. Crushed magnesite is used for the repair of furnace hearths.

Fig. 6.5 CaO-MgO phase diagram

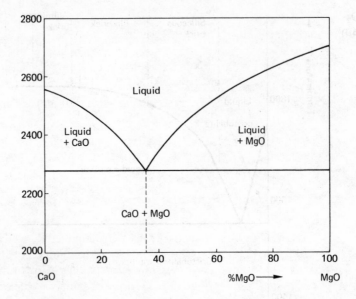

Dolomite

Good-quality dolomite is fairly plentiful in the U.K. The naturally occurring rock is basically $MgCO_3.CaCO_3$, but there is an excess of $CaCO_3$ over the true double carbonate as well as small amounts of SiO_2, Al_2O_3 and Fe_2O_3. After calcining the dolomite at high temperatures, $MgO.CaO$ is produced. The CaO-MgO phase diagram is shown in fig. 6.5.

CaO cannot be dead-burnt to become inert to moisture, so that calcined dolomite tends to crumble on exposure to moist air due to hydration of CaO. Therefore, before dolomite can be used in brick form it must be stabilised by firing with a small amount of siliceous material. The CaO present lowers the refractoriness and durability as compared with magnesite, but dolomite is a much cheaper material. A typical stabilised dolomite brick contains approximately 39% CaO, 41% MgO, 3% Fe_2O_3, 2% Al_2O_3 and 13% SiO_2.

The properties of dolomite include the following:

1 The refractoriness under load of stabilised dolomite is fairly good.
2 The resistance to basic slags is fairly good but not as good as that of magnesite.

The main uses are as a cheaper replacement for magnesite in less-critical parts of furnaces and ladles, in which the performance of dolomite may be completely satisfactory. Like magnesite it is used in crushed form (after tarring) for general furnace repair, especially for hearths.

Chrome-magnesite

The approximate composition of chrome-magnesite is 40% MgO, 30% Cr_2O_3 with some Al_2O_3, Fe_2O_3, SiO_2 and CaO.

The properties include

1 Excellent resistance to spalling.
2 Very good resistance to basic slags and to FeO.
3 Chrome-magnesite bricks are very dense but their thermal conductivity is lower than that of magnesite.
4 Medium to good resistance to abrasion and thermal shock.

Chrome-magnesite bricks are widely used in critical positions in basic-lined smelting furnaces as well as in furnace roofs. They are also used in the hearths of soaking pits of steel plants and in furnaces for re-heating of steel where attack by FeO may be severe.

Carbon

Carbon blocks are made from petroleum coke which is crushed, sized, bonded with tar, then fired. The blocks have high hot strength, high thermal conductivity, and good resistance to spalling. Molten metal and slag do not cause wetting and the blocks are resistant to chemical attack except in oxidising conditions. They can be easily machined and this enables complex shapes to be readily produced.

Carbon blocks are used in the hearth and lower part of the bosh of the iron blast furnace.

Exericses 6

1 *a*) Name two functions of a furnace refractory.
   b) Name four desirable properties of a refractory material.

2 *a*) Name three types of refractory materials.
   b) Give two examples of each type.

3 *a*) Explain what is meant by the "resistance to spalling" of a refractory material.
   b) Describe how the above property may be assessed.

4 Describe one method of determining the refractoriness-under-load of a refractory material.

5 With reference to the characteristics of carbon block refractories:
   a) name two good features, *b*) name one limitation.

6 *a*) State one way in which firebrick refractories may be classified.
   b) Name one metallurgical application of firebricks.

7 *a*) Dolomite is comprised mainly of two oxides—name them.
   b) Explain the need for "stabilisation" of dolomite.
   c) Name one advantage of dolomite over magnesite.
   d) Name one advantage of magnesite over dolomite.
   e) Name one metallurgical application of dolomite.

8 Write a short account of the polymorphism of silica.

9 Name one use for each of the following refractories:
   a) ganister, *b*) silica brick, *c*) chrome-magnesite brick.

7 Fuels and Combustion

A **fuel** may be defined as a substance which, when burned rapidly in air, evolves heat capable of being applied to industrial purposes. This definition implies that the substance should be easily ignited, burn freely, and release a large amount of heat during the complete combustion of unit quantity. The ignition point is governed by the hydrogen content of the fuel—the higher the hydrogen content the lower the ignition temperature. The burning characteristics depend on the supply of air available for combustion and on the physical and chemical properties of the fuel. The amount of heat released is mainly governed by the chemical composition of the fuel and is measured by its *calorific value*, i.e. the amount of heat released by the complete combustion of unit quantity (mass or volume) of the fuel.

7.1 Primary and Secondary Fuels

Fuels may be classified into:

1 **Primary fuels** which are naturally occurring and are obtained directly from the earth's crust or the sea.

2 **Secondary fuels** which are prepared from primary fuels (e.g. coke is made from coal).

Each of the above groups may be sub-divided into solid, liquid and gaseous types as shown below.

Primary fuels			*Secondary fuels*		
Solid	*Liquid*	*Gaseous*	*Solid*	*Liquid*	*Gaseous*
Coal	Crude petroleum	Natural gas	Coke	Fuel oil	Producer gas
				Kerosene	Water gas
					Coke oven gas
					Blast furnace gas
					Butane
					Propane

The desirable properties of a fuel include the following:

1 Easy ignition.
2 High calorific value.
3 Suitable and controllable combustion rate characteristics.
4 Low undesirable impurity content.
5 Low smoke emission.
6 Availability in quantity and uniform quality at a competitive cost.

7.2 Solid Fuels

Coal originates from decayed plant matter. It is formed by the action of high temperature and high pressure causing chemical and physical changes. The conversion sequence is as follows:

Plants→ Wood→ Peat→ Lignite→ Bituminous coal→ Anthracite

As the extent of the change increases so the carbon content increases and the hydrogen and oxygen contents decrease. Consequently the "rank" of the coal is said to increase. As the rank increases, the content of volatile matter decreases and the calorific value gradually increases.

Peat and lignite are of little or no metallurgical importance. Anthracite is a high-grade fuel which is nearly smokeless and is used industrially for steam raising. The metallurgically more important range of coals is the bituminous group which merges into the anthracites at about 90% carbon. The bituminous group can be further sub-divided as follows:

	%C	%H$_2$	%O$_2$	%Volatile matter
Long flame, non-caking	83/86	5/6	6/12	30/40
Long flame, partly-caking	83/86	4.5/5.5	5/9	30/40
Short flame, caking	85/89	4.5/5.5	4/7.5	20/30

If the combustion of a solid fuel is accompanied by flame this shows that the heat evolved is sufficient to vaporise some of the solid or that an intermediate flammable gas is being formed as a result of the oxidation of the solid. When coal is burned the flame is due to the burning of the volatile combustibles which are driven off by the heat, and also to the burning of CO formed by the incomplete combustion of carbon.

Some inorganic mineral matter exists in coal and forms the non-combustible ash residue when coal is burned; it also remains in the coke formed after coal has been carbonised (i.e. decomposed by heat, out of contact with air). Included in the mineral matter are sulphur compounds, especially pyrites and, to a much lesser extent, calcium sulphate; sulphur is also present in coal as organic compounds. Sulphur is undesirable in fuels because it is easily absorbed by solid and liquid metal, on which it usually has a harmful effect. On carbonising, some of the sulphur passes off as gaseous compounds, but usually about two-thirds or more of the sulphur present in the coal remains in the coke. When coal is burnt, a large proportion of the sulphur present is converted to SO$_2$.

The valueless, inert ash content of the coal or coke dilutes the fuel and may require extra flux and heat in certain metallurgical operations. For example, in iron smelting in the blast furnace, the fluxing of the coke is an important consideration. The amount of coarsely associated mineral matter, including pyrites, in the as-mined coal can be reduced by physical cleaning techniques such as screening. Organic compounds of sulphur, however, cannot usually be removed in this way.

Approximate *calorific values* of solid fuels are given in table 7.1.

Table 7.1 Calorific values of solid fuels

Type of solid fuel	Calorific value MJ/kg
Peat	15
Lignite	21
Semi-bituminous coal	27
Bituminous coal	31
Anthracite	35
Coke	29

The main **advantages** of solid fuel are:
1 Coal is widely distributed and is usually cheaper than other fuels.
2 Coal is the raw material for producing other products such as coke, producer gas, water-gas.

The **disadvantages** of solid fuels are:
1 They contain incombustible matter which remains after combustion as ash. This ash must be removed by raking or by fluxing to give a fluid slag.
2 They need a large amount of storage space and the storage volume is incompletely filled. Since the calorific value is also rather low, it follows that solid fuel has a low thermal storage value.
3 The properties tend to be non-uniform and successive batches can vary significantly.
4 Slow deterioration may take place with time.
5 Handling is not easy and therefore may be costly.

Production of Metallurgical Coke

Coke is made by carbonisation of a blend of several coals at about 1000°C in byproduct ovens. An oven (fig. 7.1) really consists of a battery of about 10 to 90 separate ovens. Each oven is a narrow rectangular box of silica brick 10 to 15 m long by 2 to 5 m high by 30 to 60 cm wide. Alternating with the ovens are heating chambers below which are waste-heat recovery regenerators.

Tar, benzole, naphthalene, toluene and ammonia compounds are recovered from the gas evolved during carbonisation which may be purified of H_2S before use. About 40% of the gas may be used for heating the ovens, the remainder being used for heating other plant.

The ovens are charged from the top and at the end of the coking period in a particular oven, which varies from 12 to 24 hours depending on the width, the end doors are opened and the hot coke is pushed out into a truck. The ovens are slightly tapered to facilitate the removal of the coke. The hot coke is conveyed to the quenching station where it is quenched with water and then graded according to size.

The mixture of coals used is adjusted to produce a porous, hard, strong coke. The coal is passed through a 5 mm mesh screen before being charged and each oven takes about 20 tonnes.

Fig. 7.1 Coke oven

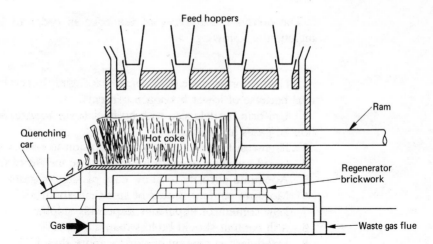

The quality of the coke produced depends on the following properties of the coal:

1   *Volatile matter content*
If this is high then the coke strength is low, whereas if the percentage volatile matter is low then the coal shows a large expansion on carbonising. A hard, strong coke is obtained with a volatile matter content of about 28–32% and blending of coals is necessary to ensure this.

2   *Ash content*
This should be less than 10% because all the ash present in the original coal mixture remains in the coke.

3   *Sulphur and phosphorus contents*
About two-thirds of the sulphur content of the coal remains in the coke, so that the initial sulphur content of the coal should be less than about 2%. All the phosphorus present in the coal remains in the coke.

Coke quality also depends on the coke oven practice so that constant attention should be paid to:

a) The blending, stocking and sizing of the raw coal.
b) Carbonising time and temperature.
c) Quenching and subsequent handling in order to avoid excessive moisture content and fines.

The required properties of the coke depend on the intended use, so that the requirements of the iron blast furnace are different from those of the iron foundry cupola.

The *functions* of coke in the smelting of iron in the blast furnace are:

a) To provide a source of heat.
b) To provide a means of reducing iron oxides to iron either directly, or indirectly.
c) To give support to the furnace burden.

The desirable *properties* of the coke in order to fulfill these functions include the following:

1 High calorific value (C.V. of coke is slightly lower than that of bituminous coal because of lower hydrogen content).
2 Uniform particle size with no fines (coke breeze) in order to allow easy flow of gases in the furnace.
3 High resistance to shatter and abrasion to resist formation of fines.
4 Good crushing strength to support the weight of the burden.
5 Adequate porosity to allow easy passage of gases.
6 Moisture content should be less than about 3%.
7 Low content of impurities, especially sulphur.
8 Ash content should be less than about 10%.
9 Fixed carbon content should be more than 85%.
10 Easily ignited (i.e. good reactivity to oxygen).
11 Low reactivity towards CO_2

7.3 Liquid Fuels

There are two main groups of liquid fuel:

1 Those derived from crude petroleum, e.g. kerosene, fuel oil (C.V. 37–45 MJ/kg).
2 Those derived from coal tar, e.g. pitch, creosote.

Crude petroleum occurs as a dark-coloured bad-smelling viscous liquid in subterranean deposits in certain parts of the world (e.g. North Sea, U.S.A., Russia, Middle East, Sahara, South America). The petroleum, which is often mixed with sand, water and brine, is extracted from its deposits by drilling deep holes. At first it issues under its own pressure, but pumping may become necessary later.

After the removal of suspended solids and dissolved gases, the crude petroleum is separated by fractional distillation into a number of portions, each corresponding to a particular range of boiling points. This is achieved by heating the crude mixture to about 400°C and passing the resulting vapours up a tall column where they condense into various portions, called *fractions*, according to their volatility. The main fuel fractions obtained are shown in table 7.2.

The **advantages** of liquid fuels are:

1 They have relatively high calorific values.
2 Because of the high C.V. and the fact that they completely fill the storage volume, they have high thermal storage values.
3 They burn without leaving a solid residue.
4 With a small volume of excess air they burn without smoke.
5 They are easy to handle and control using pumps and valves.
6 They can be stored for long periods without deterioration.

Table 7.2 Liquid fuels obtained from crude petroleum

Fraction of Crude Petroleum	B.P. range	Uses
Gaseous Hydrocarbons	below 20°C	Fuels
Naphtha { Petroleum ether	20–60°C	Solvents
Petrol (gasoline)	50–230°C	Fuel for internal combustion engines
Kerosene (paraffin oil)	200–300°C	Fuel for jet engines, tractors and stoves
Gas oil (fuel oil)	280–360°C	Fuel for domestic and industrial heating and for diesel engines (C.V. 37–45 MJ/kg)

The **disadvantages** of liquid fuels are:

1 They may be viscous and have to be heated before use.
2 They may have a relatively high sulphur content, which would be a severe disadvantage during the melting or heat treatment of many metals.

To burn fuel oil efficiently the oil must be brought into contact with the air needed for combustion at a high enough temperature and for long enough to allow complete combustion. Two methods are used to achieve this:

1 The liquid fuel is *vaporised* before ignition.
2 The fuel is *atomised* into droplets in order to expose a large surface area of the fuel to the radiant heat from the furnace and to allow it to come into contact with atmospheric oxygen. This is the method usually used in metallurgical furnaces.

Heat will be more efficiently radiated from the flame produced on combustion if the flame has a high emissivity, i.e. if the flame is luminous. The luminosity of an oil flame is due to the presence of unburnt carbon particles, which act as perfect black-body radiators of heat energy. This ability of a flame to radiate heat is very important because in metallurgical furnaces heat radiation from the flame to the material being heated is the most important method of heat transfer.

7.4 Gaseous Fuels

Gaseous fuels may be classified as indicated in table 7.3.

Natural gas is found in vast quantities underground in certain parts of the world and is often associated with known deposits of petroleum. The composition of natural gas varies with its origin, but it consists mainly of methane with small amounts of other gases. With its high C.V., ready availability, relatively low cost, cleanliness and ease of transportation and control, it is a serious rival to coal and oil as a primary fuel. Compared with some other gases it requires more air for complete combustion and its burning

Table 7.3 Classification of gaseous fuels

Gases of high calorific value	Gases of low calorific value
1. Natural gas	Gases obtained by the action of air and/or steam on carbonaceous material (coal or coke).
2. Gas obtained by the destructive distillation of coal, e.g. coke oven gas	a) Producer gas b) Water gas c) Blast furnace gas (a byproduct)

velocity is lower (because it contains no hydrogen). Consequently, it can only be efficiently burned in specially designed burners.

The averge percentage composition and calorific value of some fuel gases are given in table 7.4.

Table 7.4 Composition and calorific value of fuel gases (Bal. = balance)

Type of gas	%CO	CO_2	H_2	CH_4	N_2	Gross C.V. MJ/m^3
Natural gas (N. Sea)	—	1	—	up to 98	1	35–45
Coke oven gas	7	2	60	25	Bal.	16–22
Producer gas	25	5	15	—	Bal.	5–6
Water gas	40	5	50	—	Bal.	10–11
Blast furnace gas	30	10	1	—	Bal.	3–4

The **advantages** of gaseous fuels are:

1 They can be burned with varying amounts of air, so that a furnace atmosphere can be controlled.
2 Gases mix easily in all proportions and easy combustion is possible.
3 They are clean to use.
4 The supply to the furnace can be quickly adjusted.
5 If natural gas is used no storage is needed.
6 Gases may be liquefied to give easy storage.
7 Recovery of waste heat is easier than with other fuels.

The **disadvantages** of gaseous fuels are:

1 Safety risks regarding fire and explosion are greater than with other fuels.
2 Some gases are poisonous.
3 Leakage of gas must be guarded against.

Coke-oven gas is evolved as a byproduct during the production of metallurgical coke. It is used for heating purposes in iron and steel works, either alone or mixed with blast furnace gas.

Blast-furnace gas is a byproduct of iron smelting in the blast furnace and is

Fig. 7.2 Producer gas generator showing reaction zones

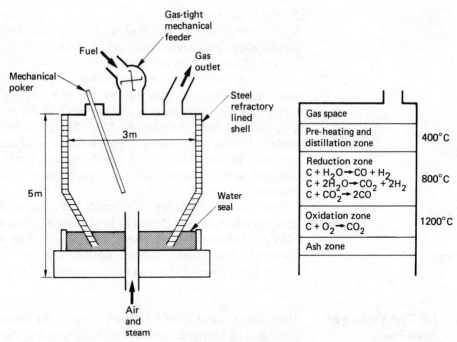

a very lean (low calorific value) fuel gas. The gas emerging from the blast furnace is laden with grit and dust so that it must be cleaned before being used to heat the hot-blast stoves. It is sometimes mixed with coke-oven gas, with resulting increase in calorific value, and then used to heat other furnaces.

Producer gas is made by the partial combustion of coal or coke in air. The gas has a fairly low C.V. but is easier to use in some processes than coke. The composition is a mixture of CO and nitrogen in the approximate ratio of 1:2 by volume—the nitrogen content accounting for the low C.V.

A producer gas generator consists of a vertical steel cylinder lined with firebrick (fig. 7.2). The whole cylinder, or part of it, may be water-jacketted. The carbonaceous solid fuel is introduced at the top, from which the producer gas is also drawn off.

The fuel bed may rest on a grate below which the air blast enters. However, there is often no real grate and the fuel bed simply rests on the ash which is gradually discharged from the bottom through a water seal. The blast inlets project into the ash and this helps to distribute the blast evenly. Large generators are usually fully automatic and work continuously.

The main chemical reactions occurring in a producer are shown in fig. 7.2, but the zones are not rigidly defined.

The composition of the producer gas generated depends on the following factors:

1 *Type of fuel used*
Bituminous coals with high contents of volatile matter give a richer gas containing small amounts of methane and tar. If coke is used, a poorer gas results with no methane or tar content.

2 *Operating temperature*

Low temperatures in the generator favour the formation of CO_2 and high temperatures favour the formation of CO.

3 *The effect of steam addition*

The introduction of steam into the blast increases the amounts of hydrogen and CO in the gas, thus raising the C.V. However, steam reduces the temperature in the generator and too much steam results in an increase in the CO_2 content, thus lowering the C.V. Therefore there is an optimum steam content for the blast. The amount of steam is adjusted to give a satisfactory fuel bed temperature, which must be sufficiently above 1000°C to give efficient gasification and yet not high enough to cause trouble by clinker formation.

The hot gas (enriched with small amounts of CH_4 and tar) from the generator may be fed directly via insulated piping to the furnace, and this "hot gas" efficiency is obviously much higher than the "cold gas" efficiency obtained when the gas is cooled prior to use.

Any H_2S present in the gas may be removed by treatment with iron oxide.

7.5 The Testing of Solid Fuels (BS 1016)

The ultimate analysis of solid fuels involves the determination of all elements present. For example, carbon and hydrogen may be determined by combustion in oxygen, forming CO_2 and H_2O which can be absorbed and weighed. Sulphur may be determined by conversion to sulphate, followed by precipitation as barium sulphate which may be ignited and weighed.

A typical analysis of a bituminous coal would be: carbon 80%, hydrogen 5%, oxygen 6%, nitrogen 1%, sulphur 1%, ash 6%.

For daily control analysis of coal, "proximate" analysis is carried out which involves standard tests for moisture, volatile matter and ash content. The moisture content should be less than about 3% and the ash content less than about 10% (the ash is mainly $SiO_2 + Al_2O_3$). "Fixed" carbon content is obtained by subtracting the sum of (the moisture + volatile matter + ash) from 100%.

Coal contains inherent moisture which is not removed when a sample is simply air-dried ready for analysis. This inherent **moisture content** is determined by heating a 1 g sample of powdered coal in a flat silica dish for 1 hour in an oven provided with air-circulation at a temperature of 107°C ± 3°C. The percentage moisture is calculated from the loss in weight.

To determine the **volatile matter content**, a 2 g sample of powdered coal, contained in a cylindrical crucible with a lid, is heated for 7 minutes at 925°C ± 15°C out of contact with air. The volatile matter is the loss in weight, less that due to moisture. The test is empirical and the value obtained depends on the temperature and time of heating. These have been fixed and, in order to obtain reproducible results, it is essential to adhere rigidly to these conditions. Air must be excluded from the coal during heating to prevent oxidation.

To determine the **ash content**, 1 g of powdered coal in a flat silica dish is heated for 2 hours in a muffle furnace at 800°C ± 10°C. The residue left behind is the ash content. The ash remaining differs in chemical composition from, and is less than, the mineral matter originally present in the coal. This is because various changes occur during the burning, e.g. loss of combined water, loss of CO_2, and oxidation of any iron pyrites present to iron oxide.

Again, the method is empirical because the conditions of incineration control the extent to which changes occur. It is therefore essential to adhere strictly to the procedure laid down in order to obtain reproducible results.

A knowledge of the **caking and swelling properties** of a coal is desirable in determining the use to which it may best be put, but there is no entirely satisfactory caking test which will enable industrial performance to be accurately predicted. Two common tests which are in use are:

a) The Gray-King Assay Test.
b) The British Standard Swelling Number Test.

In the former, 20 g of crushed coal is heated in a silica tube to 600°C and the coal visually classified A to G by comparison with standard photographs, depending on whether the residue remains a powder (A) or coalesces into a hard coherent mass without change in volume (G). Intermediate stage B is non-caking, C and D are weakly caking, while E, F and G are medium caking. Strongly caking coals swell and are designated G_1, G_2, etc. to G_{10}, the suffix indicating the number of grammes of inert carbon which must be blended into the 20 g charge to give zero swelling. The Gray-King index correlates well with coal rank and volatile matter and is probably the best index of caking capacity of a coal or coal blend.

In the British Standard test, 1 g of powdered coal is heated at a standard rate in a crucible with a lid until volatile matter ceases to be evolved and the size and shape of the residue are compared with standard charts.

The **calorific value** of *solid* and also *liquid* fuels can be determined by burning a known mass in oxygen at a pressure of not less than 25 atmospheres. The combustion is carried out in a steel bomb which stands in a calorimeter vessel containing a known mass of water and whose temperature can be accurately measured (fig. 7.3). The calorimeter is enclosed in a jacket

Fig. 7.3 Bomb calorimeter

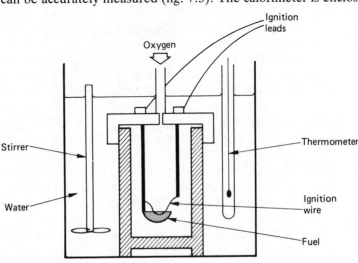

to minimise heat interchange with the surroundings. Provided that the temperature rise in the water and the water equivalent of the apparatus are accurately known, then the calorific value can be determined. A small correction for radiation heat loss is included in the calculation for most accurate results.

7.6 The Testing of Liquid Fuels

Fuel oils may be subjected to a variety of tests depending on their application.

Density, and its variation with temperature, may be required to be known for conversion of volume to mass. **Flash point** determination (BS 2839) gives information about handling and storage dangers.

The flash point is the temperature of the oil at which a flash first appears at any point on the surface of the oil when a standard test flame is applied. Minimum flash points are stipulated by law for various grades of oil (e.g. >80°C for fuel oil). Flash points may be determined in standard equipment such as the Pensky-Martens apparatus (fig. 7.4). The test is carried out mainly as a safety precaution, but would also indicate deviation from specification.

The **fire point** may be defined as the temperature at which the oil ignites and continues to burn for 5 seconds.

Viscosity is determined as a check on specification because consistent behaviour in burners and pipelines is desirable. Viscosity varies logarithmically with temperature and is best determined over a range of temperature.

A commonly used apparatus is the Redwood Viscometer (fig. 7.5) in which two sizes of orifice are available. The apparatus comprises an oil cup in a

Fig. 7.4 Pensky-Martens flashpoint apparatus

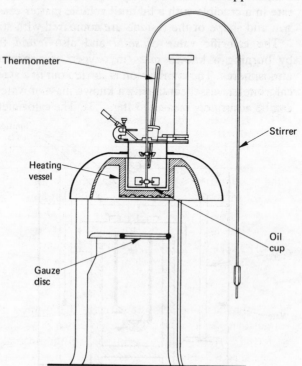

Thermometer

Stirrer

Heating vessel

Oil cup

Gauze disc

Fig. 7.5 Redwood
no. 1 viscometer

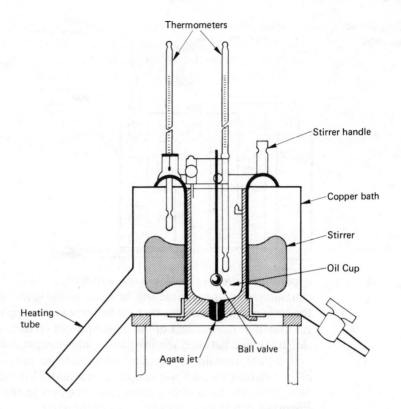

Thermometers

Stirrer handle

Copper bath

Stirrer

Oil Cup

Heating
tube

Agate jet

Ball valve

water jacket which can be temperature-controlled. When conditions are stabilised, the oil is allowed to flow through a standard orifice in the base of the cup and the viscosity is expressed as the number of seconds for 50 cm³ of oil to flow. When the flow time in Viscometer I (with small orifice) exceeds 2000 secs ($\simeq$ 33 mins), Viscometer II may be used, in which the orifice is larger and the time is reduced by a factor of 10.

Viscosities of fuel oils lie in the range 250–1000 Redwood I secs at about 40°C. Oil can be pumped if the viscosity is less than about 2000 Redwood I secs, but industrial burners need a viscosity of 100–150 Redwood I secs, so that the oil must be warmed and this is usually done by steam pipes.

Fuel oils may contain about 2–4% sulphur. The **sulphur content** may be determined by burning a known mass of oil in oxygen and measuring the amount of SO_2 evolved. Sulphur is a very objectionable impurity because it can cause embrittlement of metallic materials and also result in stained surfaces during heat treatment.

7.7 The Testing of Gaseous Fuels

The **calorific value** of gaseous fuel is determined by burning a known volume of gas in a steady-flow calorimeter, e.g. a Boys Calorimeter (fig. 7.6). The gas is fed at a steady rate to a burner and the products of combustion pass over tubes through which there is a steady flow of water. When all parts of the

Fig. 7.6 Diagram of Boys calorimeter

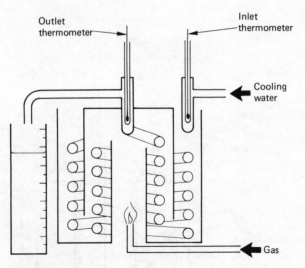

apparatus have reached a constant temperature, the heat released by combustion in a given time will be equal to the heat taken in by water during the same period (assuming no heat losses from the apparatus), and this can be calculated as the product of the mass of water collected, its specific heat, and the difference between the inlet and outlet temperatures.

If a fuel contains hydrogen, water will be produced during the C.V. determination and this will condense to liquid. However, in the practical use of such fuels, the water in most cases remains in the gaseous state (i.e. as steam) so that the latent heat of combustion of the water is not liberated, and the full C.V. of the fuel is not attained. Therefore, it is usual to distinguish between the *gross* C.V. as determined in the calorimeter, and the *net* C.V., which is the value after deduction of the latent heat given up by the condensation and cooling of any water present in the products of combustion (including any water present originally in the fuel).

It is not usually possible to cool down the waste gases to liberate the latent heat of condensation. Also, most flue gases pass out of a furnace at temperatures much higher than 100°C and they carry away not only the latent heat of the uncondensed steam, but also additional heat from the furnace.

Samples of the same fuel are likely to give different net calorific values under different working conditions, so that when comparing different fuels the *gross* calorific value is the proper value to take.

Gas analysis matters which are of significance to the fuel technologist are:

1   The analysis of the flue gas
2   The analysis of fuel gas.

The routine **analysis of flue gas** is carried out using the Orsat apparatus (fig. 7.7). The principle of the apparatus involves absorption from the mixture of the gas sought by a reagent, followed by measurement of the reduction in volume. Carbon dioxide, oxygen and carbon monoxide are usually estimated by absorption in caustic potash, alkaline pyrogallol and ammoniacal cuprous chloride respectively.

Fig. 7.7 Orsat gas analysis apparatus

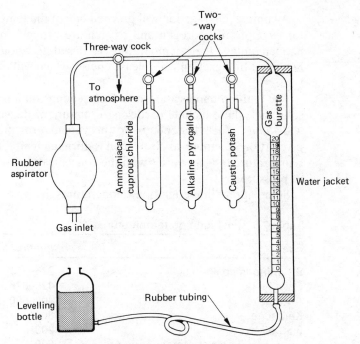

Continuous assessment of flue gas may be desirable and this is achieved by attaching recorders to permanent sampling lines in the flues. Carbon dioxide recorders, for example, use the principles of either absorption in potash or thermal conductivity of gases. Carbon dioxide has a thermal conductivity about half that of oxygen and nitrogen, so that if a sample of dry flue gas is passed over one arm of a heated Wheatstone bridge circuit the out-of-balance produced will be a measure of the carbon dioxide. Apparatus is also available for recording the oxygen content of a flue gas—most industrial recorders using the magnetic properties of oxygen for its measurement.

The **analysis of fuel gas** is expressed as percentages by volume of the molecular species present. A volume of gas is brought into contact with a series of reagents which absorb the constituent gases one at a time. Carbon dioxide, oxygen and carbon monoxide absorbtion is carried out as in the Orsat method. Unsaturated hydrocarbons can be absorbed in fuming H_2SO_4. Combustible gases such as hydrogen and saturated hydrocarbons cannot be absorbed in this way and are determined by combustion or explosion with excess oxygen, the decrease in volume being measured before and after absorbing any carbon dioxide produced.

7.8 Combustion of Fuels

The **combustion of a fuel** is a chemical reaction in which certain constituents combine with oxygen (usually provided by the atmosphere) and is accompanied by a release of energy in the form of heat. The combustible elements are carbon and hydrogen and these make up the bulk of all fuels. (In some metallurgical materials there may be present a considerable amount of sulphur, which is also a combustible element.)

A combustion reaction will proceed only if the temperature is high enough; for example, if hydrogen and oxygen are brought into contact nothing will happen unless their temperature is raised to about 600°C. Once started however, the reaction will release energy which will maintain a high temperature.

The **ignition temperature** of a fuel depends on a number of factors. For a solid fuel, these include particle size, nature of the surface, and its porosity; with *liquid* fuels it depends on the vapour pressure of the liquid and the proportion of vapour to air needed to form an ignitable mixture. The ignition temperature of a *gas* varies with the concentration and is lowered by increase in pressure.

Some approximate ignition temperatures are given in table 7.5.

Table 7.5 Fuel ignition temperatures

Fuel	Ignition temperature (approx.) °C
Bituminous coal	275–325
Coke	425–625
Anthracite	400–500
Kerosene	300
Fuel oil	200
Hydrogen (in air at atmospheric pressure)	600
Carbon monoxide	650
Methane	700

Combustion may take place steadily over a long time, as in a furnace, or it may be an almost instantaneous process, as in the cylinder of an internal combustion engine. In all cases, the three essentials of the combustion process are a supply of fuel, a supply of oxygen or air, and a means of ignition, i.e. a way of producing a temperature high enough to start the reaction.

Calorific value can be calculated from the chemical composition if all the heats of reactions occurring during combustion are known. Certain assumptions have to be made, notably that combustion is complete to CO_2 and H_2O, which often does not happen in a furnace.

Calculation of C.V. is most successful in the case of gaseous fuels, where compounds are relatively simple and their heats of combustion are usually known. The accuracy of the determination then depends on that of the gas analysis.

Calorific Value Calculations

Example 1   A bomb calorimeter was used to determine the calorific value of a sample of coal and the following observations were made:

mass of coal ignited 1.25 g; mass of water in container 2.0 kg;
temperature before ignition 17.575°C;
the maximum temperature reached 21.075°C.

The water equivalent of the apparatus was 0.50 kg. Assuming no heat loss from the apparatus, and taking the specific heat of water as 4.18 kJ/kg°C, calculate the C.V. of the coal.

Heat released by combustion of coal
= Heat taken in by water and apparatus
Let x = C.V. of the coal. Then
$$x \times 1.25 \times 10^{-3} = (2.0 + 0.5) \times 4.18 \times (21.075 - 17.575)$$
$$x = \frac{2.5 \times 4.18 \times 3.50}{1.25 \times 10^{-3}}$$
$$= 29.26 \times 10^3 \, \text{kJ/kg} = 29.3 \, \text{MJ/kg}$$

Example 2 In a test on a gas using a steady flow calorimeter, the following results were obtained: volume of gas burned 0.02 m³; temperature of gas 20°C; gauge pressure of gas (measured by a water manometer) 50 mm water; mass of water passed through calorimeter 4.5 kg; inlet water temperature 10.25°C; outlet water temperature 28.75°C; barometric pressure 101.5 kN/m². Taking the specific heat of water as 4.18 kJ/kg°C and its density as 1000 kg/m³, calculate the calorific value of the gas in MJ/m³ at s.t.p. (s.t.p. is 0°C and 101.325 kN/m²) assuming no heat loss from the apparatus. (Take acceleration due to gravity, g, as 9.81 m/s².)

Heat taken in by cooling water
= 4.5 × 4.18 × (28.75 − 10.25) kJ
= 4.5 × 4.18 × 18.5 kJ
= 348.0 kJ

Manometer pressure of gas
= height of water column × density × g
= 0.05 × 1000 × 9.81 N/m²
= 490.5 N/m²

Absolute pressure of gas
= (490.5 + 101 500) N/m²
= 101 990.5 N/m²
= 101.9905 kN/m²

The volume of gas used at s.t.p. is found by applying the gas law:
$$p_1V_1/T_1 = p_2V_2/T_2 \qquad V_2 = p_1V_1T_2/p_2T_1$$
∴ Required volume of gas
$$= \frac{101.9905 \times 0.02 \times 273}{101.325 \times 293} \, \text{m}^3 = 0.018\,76 \, \text{m}^3$$

Let x = C.V. of gas, then
$$x \times 0.018\,76 = 348.0 \, \text{kJ/m}^3$$
$$x = 348.0/0.018\,76 \, \text{kJ/m}^3$$
$$= 18\,550 \, \text{kJ/m}^3 = 18.55 \, \text{MJ/m}^3 \text{ at s.t.p.}$$

Combustion of Chemical Elements

The chemical elements concerned in combustion processes are listed below:

Element	Relative atomic mass
Hydrogen	1
Carbon	12
Sulphur	32
Oxygen	16
Nitrogen	14

The combustion of a fuel involves the chemical combination of its constituents with oxygen. If the combustion is complete, the final products will be CO_2, H_2O and, if sulphur is present, SO_2. It is seen therefore that the result of combustion is the same as if the carbon, hydrogen and sulphur had burned separately. Hence, for combustion purposes each element may be dealt with independently. Although sulphur is present in many fuels, the amounts usually involved are very small and in calculations concerned with air requirements are usually ignored.

Combustion of Carbon

The complete combustion of carbon is represented by

$$C + O_2 \rightarrow CO_2$$

i.e. 1 molecule of carbon (consisting of 1 atom) and 1 molecule of oxygen (consisting of 2 atoms) react to form 1 molecule of carbon dioxide. Remembering that there are 6.022×10^{26} molecules in $1\,k\,mol$ (Avogadro's constant), then considering this number of molecules instead of 1 molecule, the amount of each substance becomes $1\,k\,mol$ (defined as the mass in kg numerically equal to the relative molecular mass of the substance).

$$\therefore \quad 12\,kg \text{ carbon} + 32\,kg \text{ oxygen} \rightarrow 44\,kg \text{ carbon dioxide}$$

Combustion of Hydrogen

The combustion of hydrogen may be dealt with similarly:

$$2H_2 + O_2 \rightarrow 2H_2O$$
$\therefore$ $2\,k\,mol$ hydrogen + $1\,k\,mol$ oxygen $\rightarrow$ $2\,k\,mol$ water
or $4\,kg$ hydrogen + $32\,kg$ oxygen $\rightarrow$ $36\,kg$ water

Under the same conditions of temperature and pressure, $1\,k\,mol$ of any gas occupies the same volume. Therefore the *volume* in the previous equation will (assuming the H_2O to be a gas, i.e. superheated steam) be in the proportions

2 volumes hydrogen + 1 volume oxygen $\rightarrow$ 2 volumes steam

The total *mass* after the reaction is equal to the total *mass* before the reaction, but this rule does *not* apply to *volumes*.

Air Required for Combustion

Air is a mixture of gases and an accurate analysis by mass is 23.2% oxygen, 75.5% nitrogen and 1.3% made up of rare gases such as argon and neon. It also contains a very small amount of carbon dioxide and varying amounts of water vapour. The carbon dioxide, water vapour and the rare gases are ignored in combustion calculations and the whole of the air other than oxygen is considered as "atmospheric nitrogen". Therefore, for practical purposes, air is considered to consist by *mass* of 23% oxygen and 77% nitrogen; and by *volume* of 21% oxygen and 79% nitrogen.

The amount of oxygen required for the combustion of the carbon and hydrogen can be found by using the equations given above. The fuel itself may contain oxygen and, if so, the amount of oxygen in the fuel is subtracted from that required for combustion to give the oxygen to be supplied by the air. Since air contains 23% oxygen by mass, this quantity of oxygen must be multiplied by 100/23 to give the mass of air required; this is the *minimum* amount of air for complete combustion of the fuel, and is known as the "*theoretical air*" requirement. If the amount of air supplied to a furnace is the "theoretical air", it is unlikely that complete combustion will result. The fuel and oxygen will both be present in the correct proportion, but it is unlikely that they will react completely. In practice, some molecules of both fuel and oxygen will pass through the combustion zone without having come into contact. To avoid the resulting wastage of fuel it is usual to supply more air than the theoretical amount, and the additional amount of air supplied is referred to as the "*excess air*". The percentage of excess air supplied varies with the design of the furnace and on the fuel used. Gaseous fuels need about 10% excess air, while liquid and solid fuels require about 15% and about 50% respectively.

The air brought into the combustion zone upstream of the fuel is called *primary* air.

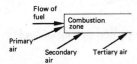

As it passes through the fuel it loses oxygen, and on the downstream side of the combustion zone, the shortage of oxygen may cause the endothermic reaction $CO_2 + C \rightarrow 2CO$ to set in. To prevent this, *secondary* air is introduced into the combustion zone at this point. A third entry point may be required with some fuels, e.g. with long-flamed bituminous coal, in order to burn volatilised hydrocarbons.

Combustion Products

If the combustion is incomplete, some of the carbon atoms will combine with only one oxygen atom to form carbon monoxide CO instead of CO_2. The products of combustion may also include unburned hydrocarbons and unburned carbon in the form of "smoke".

The composition of the flue gases from a furnace is thus an indication of whether the combustion is complete, and may also be used to determine whether the correct amount of excess air is being supplied. Products of combustion may be analysed by various means and, if the analysis is given by volume, it will not include water vapour because after condensation the volume of water becomes negligible.

Combustion calculations are, therefore, often concerned with the assessment of the percentage composition of the dry products of combustion. Compositions may be calculated by mass or by volume; the latter enables comparison to be made with volumetric gas analysis.

Combustion Calculations—Air Requirement

Example 1 Find the minimum mass of air necessary for the complete combustion of 1 kg of coal with the following analysis by mass: carbon 81%, hydrogen 5%, oxygen 6%, ash 8%. If 40% excess air is supplied find the percentage composition of the dry products of combustion *a*) by mass, *b*) by volume.

1 kg of coal contains 0.81 kg carbon, 0.05 kg hydrogen and 0.06 kg oxygen.

$$C + O_2 \rightarrow CO_2$$
By mass $\quad$ 12 kg + 32 kg $\rightarrow$ 44 kg

$\therefore \quad$ 0.81 kg carbon requires $0.81 \times 32/12$ kg oxygen = 2.16 kg

$$2H_2 + O_2 \rightarrow 2H_2O$$
By mass $\quad$ 4 kg + 32 kg $\rightarrow$ 36 kg

$\therefore \quad$ 0.05 kg hydrogen requires $0.05 \times 32/4$ kg oxygen = 0.4 kg

$\therefore \quad$ Total mass of oxygen required = 2.16 + 0.4 = 2.56 kg

The coal contains 0.06 kg oxygen.

$\therefore \quad$ Mass of oxygen to be supplied = 2.56 − 0.06 = 2.5 kg

Air contains 23% oxygen by mass.

$\therefore \quad$ Minimum mass of air required $= 2.5 \times \dfrac{100}{23} = 10.87$ kg

40% excess air means that the actual amount of air supplied per kg of fuel is

$$10.87 \times 140/100 \text{ kg} = 15.22 \text{ kg}$$

and this will consist of

oxygen $\quad$ $15.22 \times 0.23 = 3.50$ kg $\quad$ and the remainder
nitrogen $\quad$ $15.22 - 3.50 = 11.72$ kg

Because the composition of the *dry* products of combustion are required, the water formed can be ignored.

$$CO_2 \text{ produced} = 0.81 \times 44/12 = 2.97 \text{ kg}$$

The dry products of combustion will consist of this carbon dioxide together with the excess oxygen 3.50 − 2.50 = 1.0 kg and the whole of the nitrogen 11.72 kg.

Thus the composition, by mass, of the products is

CO_2	2.97 kg	%CO_2	= 2.97/15.69 × 100 =	18.9
O_2	1.00 kg	%O_2	= 1.00/15.69 × 100 =	6.4
N_2	11.72 kg	%N_2	= 11.72/15.69 × 100 =	74.7
Total	15.69 kg			100.0

To find the composition of the combustion products by *volume*, it must be remembered that in a mixture of gases the proportions by volume are the same as proportions by k mol. In the combustion of 1 kg of fuel, the composition by k mol of the dry products of combustion is:

CO_2	2:97/44	= 0.0675 k mol
O_2	1.00/32	= 0.0313 k mol
N_2	11.72/28	= 0.4185 k mol

If, at the temperature and pressure of the combustion products, 1 kg of any gas occupies a volume V, the composition of the dry products of combustion by volume is therefore:

CO_2	0.0675 V	%CO_2	= 0.0675/0.5173 =	13.1
O_2	0.0313 V	%O_2	= 0.0313/0.5173 =	6.0
N_2	0.4185 V	%N_2	= 0.4185/0.5713 =	80.9
Total	0.5173 V			100.0

Example 2 The volumetric composition of a fuel gas is as follows: 85% methane, 5% hydrogen and 10% nitrogen.

Find *a*) the minimum volume of air required for the complete combustion of 1 m^3 of gas, *b*) the volumetric analysis of the dry products of combustion if minimum air is supplied and combustion is complete.

For combustion of methane:

$$CH_4 + 2O_2 \rightarrow CO_2 + 2H_2O$$

Proportions by volume are the same as proportions by k mol, i.e.

1 vol + 2 vols → 1 vol + 2 vols

∴ 0.85 m^3 methane require 2 × 0.85 = 1.70 m^3 oxygen

and the dry product of its combustion is 0.85 cm^3 carbon dioxide.

For combustion of hydrogen:

$$2H_2 + O_2 \rightarrow 2H_2O$$

i.e. 2 vols + 1 vol → 2 vols

0.05 m^3 hydrogen require $\frac{1}{2}$ × 0.05 = 0.025 m^3 oxygen

In this case there is no "dry product of combustion".

$$\therefore \quad \text{Total volume of oxygen required} = (1.70 + 0.025)\,\text{m}^3$$
$$= 1.725\,\text{m}^3$$

Air contains 21% oxygen by volume, therefore the minimum air requirement is

$$1.725 \times 100/21 = 8.214\,\text{m}^3$$

Combustion Products

Since minimum air is used, there will be no oxygen in the combustion products, so that the dry products of combustion will be nitrogen from the air and from the fuel, and CO_2 from the combustion of the methane.

$$\text{Volume of nitrogen in air supplied} = 8.214 \times 0.79 = 6.49\,\text{m}^3$$

Fuel gas contains $0.10\,\text{m}^3$ nitrogen.

$$\therefore \quad \text{Total volume of nitrogen} = 6.49 + 0.10 = 6.59\,\text{m}^3$$

Therefore the composition of the dry products of combustion is as follows:

N_2	$6.59\,\text{m}^3$	$\%N_2$ $= 6.59/7.44 \times 100 =$	88.6
CO_2	$0.85\,\text{m}^3$	$\%CO_2 = 0.85/7.44 \times 100 =$	11.4
Total	$7.44\,\text{m}^3$		100.0

Example A fuel gas consists, by volume, of 88% methane, 4% hydrogen and 8% nitrogen. Find the volumetric composition of the dry products of combustion if $1\,\text{m}^3$ of gas is burned with $10\,\text{m}^3$ of air.

For combustion of methane $\quad CH_4 + 2O_2 \rightarrow CO_2 + 2H_2O$
$\therefore \quad 0.88\,\text{m}^3$ methane requires $2 \times 0.88\,\text{m}^3$ oxygen $= 1.76\,\text{m}^3$ and dry product combustion is $0.88\,\text{m}^3$ carbon dioxide.
  For combustion of hydrogen $\quad 2H_2 + O_2 \rightarrow 2H_2O$
$\therefore \quad 0.04\,\text{m}^3$ hydrogen requires $\frac{1}{2} \times 0.04\,\text{m}^3$ oxygen $= 0.02\,\text{m}^3$ and there is no dry product of combustion.
$\therefore \quad$ Total volume of oxygen required $= 1.76 + 0.02 = 1.78\,\text{m}^3$
  Total volume of air used $= 10\,\text{m}^3 = 2.1\,\text{m}^3$ oxygen $+ 7.9\,\text{m}^3$ nitrogen
$\therefore \quad$ Excess oxygen $= 2.1 - 1.78 = 0.32\,\text{m}^3$
  Volume of nitrogen in fuel gas $= 0.08\,\text{m}^3$
$\therefore \quad$ Total nitrogen $= 0.08 + 7.9 = 7.98\,\text{m}^3$

Composition of dry combustion products is

CO_2	$0.88\,\text{m}^3$	$\%CO_2 = 0.88/9.18 \times 100 =$	9.6
O_2	0.32	$\%O_2$ $= 0.32/9.18 \times 100 =$	3.5
N_2	7.98	$\%N_2$ $= 7.98/9.18 \times 100 =$	86.9
Total	9.18		100.0

Exercises 7

1   *a*) Distinguish between a primary and secondary fuel.
    b) Give two examples each of primary and secondary fuels.
    c) Name four desirable technical characteristics of a fuel.

2   *a*) Describe a test used for assessing the caking properties of a coal.
    b) Name four desirable properties required of blast furnace coke.

3   *a*) State the properties determined in the proximate analysis of coal.
    b) Describe the determination of one property mentioned in (*a*).

4   2 g of coal were placed into a weighed silica container which was heated to 110°C,
    allowed to cool and weighed (*a*). The container was then partially covered with a lid
    and heated for 7 minutes at 925°C, allowed to cool and re-weighed (*b*). Finally, the
    container was strongly heated without the lid until all the specks of carbon had
    disappeared and the residue was cooled and the container re-weighed (*c*). The
    following results were obtained:

    Mass of empty container   = 16.780 g
    Mass of container after heating (*a*) = 18.735 g
                                     (*b*) = 18.360 g
                                     (*c*) = 16.890 g

    Calculate the percentage composition of the following: moisture , volatile matter,
    ash and fixed carbon.

    [*Ans.* 2.5%, 32.5%, 8.5%, 56.5%]

5   *a*) Name two types of liquid fuel.
    b) Name two advantages and two disadvantages of liquid fuel compared with solid
    fuel.
    c) Define the term "flash-point" and describe a method for its determination.

6   A bomb calorimeter was used to determine the calorific value of a sample of oil and
    the following results were obtained: mass of oil in crucible 1.025 g; mass of water in
    container 2.5 kg; temperature before ignition of fuel 17.525°C; maximum temperature
    reached 20.925°C. The water equivalent of the apparatus was 0.75 kg. Assuming no
    heat loss from the apparatus, and taking the specific heat of water as 4.18 kJ/kg°C,
    calculate the calorific value of the oil.

    [*Ans.* 45.1 MJ/kg]

7   *a*) State the approximate composition of three gaseous fuels.
    b) Name four advantages of gaseous fuels.
    c) Name two hazards associated with the use of gaseous fuels.
    d) A sample of natural gas consists of 90% methane and 10% hydrogen.
    Find the minimum volume of air required for the complete combustion of $1 m^3$ of
    the gas. (Assume that air contains 21% oxygen by volume.)

    [*Ans.* $8.81 m^3$]

8   Calculate the minimum mass of air required for the complete combustion of 1 kg of oil
    of the following composition by mass: carbon 86%, hydrogen 14%.
    The oil is burned using 20% excess air. Find the percentage composition of the dry
    products of combustion *a*) by mass, *b*) by volume. (Air contains 23% of oxygen by
    mass. Relative atomic masses are as follows: carbon 32, hydrogen 1, oxygen 16,
    nitrogen 14.)

    [*Ans.* 14.83 kg; *a*) CO_2 18.0%, O_2 3.9%, N_2 78.1%
    b) CO_2 12.3%, O_2 3.7%, N_2 84.0%]

8 The Joining and Machinability of Metals

8.1 Joining Processes

There are innumerable instances in which metals have to be joined to one another. In many cases joining is part of the production process, e.g. a component may be too complicated or too big to make in one piece and several simpler or smaller parts may have to be joined together.

Joining processes may be classified as follows:

1 Mechanical joining
2 Adhesive bonding
3 Soldering and brazing
4 Welding

Industrial applications of these processes are many and varied ranging from the soldering of cans to the welding of oil rigs.

8.2 Mechanical Joining

There are many types of mechanical joint, some of which are only required to be temporary, e.g. to allow the assembly to be broken down for adjustment or maintenance. Mechanical fixings may be classified as follows:

1 Rivets
2 Nuts, bolts, screws and studs
3 Pins
4 Spring washers and retaining rings
5 Quick-release fasteners.

A **rivet** needs a preformed hole of suitable size and, after insertion, the rivet must have a head worked on, similar in proportion to the forged head on the other side. In addition to solid rivets, tubular, and bifurcated rivets (fig. 8.1) are also widely used. When joints are required to be airtight or watertight (as with boiler plates or ship's plates), the rivet is first heated to a red heat. On cooling, the rivet contracts and a very tight joint is produced.

Fig. 8.1 Types of rivet

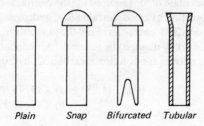

Plain   Snap   Bifurcated   Tubular

Nuts and bolts are very widely used as fastening devices. Clearance holes are required, but no thread is needed inside the component. Bolts are threaded for only part of their length, but both the nut and the bolt head must be readily accessible for tightening with a spanner.

There are many cases when it is not possible to use a nut and bolt (e.g. lack of accessibility) and a **set screw** is then used. Holes inside the components to be joined are threaded for the greater part of their length and the screw is then sunk below the surface.

Studs are threaded at both ends and are used when heavy tightening forces are needed. Tightening of the nut placed on the head is best achieved by a torque wrench in order to prevent over-tightening which can cause failure.

All the above threaded fasteners provide the necessary forces that pull together the parts to be joined.

Pins are commonly used for simple insertion into drilled holes with minimum clearance. Split pins are locked by opening the protruding ends, while self-locking types are also used.

Spring washers or *retaining rings* are easily placed in position and may be located in grooves. They are used to assist in locking nuts and bolts against loosening due to vibration.

Quick-release fasteners are available in a wide range of types and are mainly used for temporary joints and to hold panels or decorative trim in position.

8.3 Adhesives

Very powerful **adhesives** are now available for engineering applications. These materials are synthetic and need a curing treatment; some resins are self-curing, while others require heating. Heat may be applied only to the required areas by means of radio-frequency waves. The design requirements regarding joint strength, service temperature and resistance to moisture must be taken into account when choosing an adhesive, as well as the relevant properties of the adhesive, including storage-life, curing treatment, odour, toxicity, and method of preparation.

Joint design is an important factor in achieving a strong bond, and in general the bond strength is proportional to the area bonded. The lap joint (fig. 8.2b) is the most widely used joint because of its large bonding area and simplicity. As with all joining processes, the condition of the surfaces to be joined is most important and very careful preparation is vital. The surfaces are roughened with abrasive and then thoroughly cleaned to remove all grease and other impurities. The adhesive is then applied and the joint carefully made by placing together the two surfaces requiring bonding. Jigs

Fig. 8.2 Types of joint

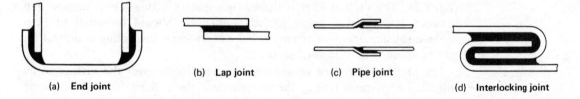

(a)　**End joint**　　(b)　**Lap joint**　　(c)　**Pipe joint**　　(d)　**Interlocking joint**

and fixtures are often used to line up the component parts because bonding may take place instantly.

A wide variety of adhesives is available commercially. Epoxy resins are widely used since they will bond to almost any surface. They can be cured to give a cross-linked thermosetting matrix by using a catalyst (hardener) addition or by heating. The joints produced are strong in tension and shear but tend to be brittle. This brittleness can be overcome by making modifying additions such as nylon (a thermoplastic) to the resin. This increases flexibility of the adhesive and improves the ability to "wet" (i.e. spread over) the surfaces of impervious materials such as glass and metal.

Recent developments include chemically blocked adhesives, the commonest example of which is Superglue. This is a cyanoacrylate which cures when in contact with the water vapour in the atmosphere. They are, however, expensive and their very short curing times restricts their application. However, they are useful for holding components temporarily in position for processes such as welding.

Other joining processes described below (soldering, brazing, welding) involve treatment at higher temperatures which produce structural changes in the parent metal, as well as causing localised dimensional changes resulting in distortion and internal stresses. Joining by adhesives does not suffer from these disadvantages.

8.4 Soldering and Brazing

In soldering and brazing, although the surfaces to be joined are heated the temperatures developed remain well below the melting point of either of the materials to be joined. However, local dimensional changes may cause distortion of the workpieces and result in residual stresses. If excessive heating is allowed to occur, then structural changes may take place in the materials.

A *filler alloy* is used which has a much lower melting point than the fusion temperature of the materials being joined. If the filler alloy melts above 500°C, the joining process is called *brazing* (or hard soldering), while *soldering* (also called soft soldering) processes take place below about 400°C. At these relatively low temperatures only a small amount of atomic diffusion is possible so that joint strengths are rather weak. However, brazed joints are significantly stronger (about 350–450 N/mm^2) in tension than soldered joints (about 50–60 N/mm^2) because of the greater diffusion, and also because brazing alloys are stronger than solders.

Another factor which greatly influences joint strength is the design of the joint. In the case of joints which are only very lightly stressed, lap joints (end joints) are used, while for greater strength an interlocking joint is preferable (fig. 8.2). The molten filler is drawn by capillary action into a narrow gap between the surfaces being joined. Joint gaps should be small to help achieve satisfactory joint strength, but not so small that filling is difficult; a gap of about 0.125 mm is often used.

The molten filler must be capable of spreading all over the surface to be joined, i.e. it must "wet" the surface and there must be at least some tendency for alloying to take place. This means that the surface must be perfectly clean and this may be achieved by a combination of chemical and mechanical methods. Acid pickling may be used to remove surface oxide but care must be taken to ensure complete removal of any remaining acid; oil and grease may be removed by solvent degreasing, while final cleaning may be by abrasion. In addition to the thorough cleaning, the use of an active flux is also necessary to act as a solvent for oxide films formed during the joining operation. If the flux is corrosive, any residual material must be removed on completion of joining.

Heat may be supplied in a variety of ways including soldering iron, gas torch, electric hot plate, oven, induction, salt bath, controlled atmosphere furnace, and vacuum furnace. Both soldering and brazing operations may be carried out manually or by semi-automatic equipment.

Soldering

Materials used for soft soldering are covered by BS 219, and consist mainly of lead-tin alloys which melt within the range 175–300°C (fig. 8.3). An addition of up to about 3% antimony may be made to replace the expensive tin, but these solders are not suitable for joining galvanised components because of the formation of brittle compounds with zinc. Soft solders are available as cast bar, extruded section, wire, flux-cored wire and as a paint.

Tinman's solder (62% Sn, 38% Pb) is of exactly eutectic composition and therefore freezes quickly; it is used for a wide range of electrical and electronic components including instrument work, radio and television sets. In electrical work the conductivity of the joint is obviously important and, since an exceedingly large number of joints are usually involved, close control of the soldering operation is necessary.

Fig. 8.3 Lead-tin equilibrium diagram

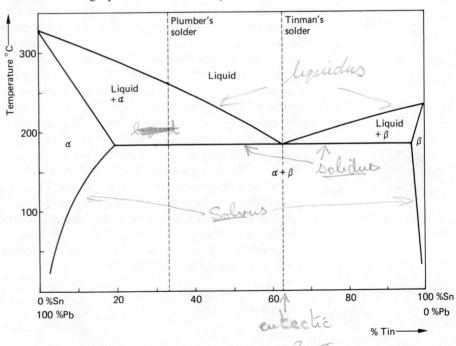

An alloy containing 45% Sn, 55% Pb is used in general engineering work, while Plumber's solder (33% Sn, 67% Pb) is used for pipe joints. The latter alloy has a pasty stage during its freezing—hence the "wiped" joint in plumber's work.

As mentioned previously, soldered joints are rather weak (especially in tension), so that the interlocking type of joint is often used to improve the strength.

A *flux* is applied to the joint either before soldering or during the soldering operation in the case of flux-cored wire. The flux exposes clean metal surfaces and prevents oxidation during the heating cycle. The flux also assists the "wetting" process and filling of the joint. Typical fluxes include zinc chloride solution ("killed spirits") and ammonium chloride solution, but these are corrosive and must be removed by washing after soldering. For electrical work, resin-cored solder wire is usually used, thus avoiding corrosion problems. The use of a flux is, of course, unnecessary when vacuum heating is used. The solder may be applied manually or placed in position before heating.

Brazing

Alloys used for brazing are dealt with in BS 1723 and BS 1845, and melt in the approximate temperature range 500–900°C. Because of the higher temperatures involved in brazing, more atomic diffusion occurs, causing alloying between the brazing alloy (spelter) and the base material and resulting in stronger joints than in soldering, so that the use of interlocking joints is unnecessary. The greater diffusion also leads to better "wetting" of the surface by the molten filler. Thorough cleaning of the surface is again essential, as also is the use of a flux. If brazing is carried out at temperatures below about 750°C a fluoride flux may be used, but above 750°C borax (sodium tetraborate) is usually used. During torch brazing, the parts to be joined are fluxed; then the filler is applied manually from pre-fluxed alloy wire. In other heating methods, the brazing alloy in the form of paste, strip, disc or coil of wire may be positioned at the joint area, then fluxed and heated to the required temperature. After the brazing operation, residual flux must be removed, e.g. by a water rinse. The use of a controlled atmosphere furnace or vacuum heating may obviate the need for a flux.

The oldest type of brazing alloy in use is brass containing 50–60% copper, which melts at about 850–900°C and is widely used for joining ferrous materials. For general work, alloys of Cu, Zn, Ag and Cd are often used. These alloys are known as hard solders or silver solders and melt in the range 600–800°C; they are of approximately eutectic composition and flow readily. Furthermore, they form stronger joints than brass.

For joining aluminium-base alloys, an alloy containing 86% Al, 10% Si and 4% Cu is used, which melts in the range 550–600°C.

8.5 Welding Processes

Welding processes may be divided into two main groups depending on whether or not pressure is used to bring the pieces to be joined together. In **fusion welding** without pressure, the faces of the two pieces of metal to be joined are heated and melted and extra molten metal from a filler rod may be added. The filler metal is of the same or similar composition to the metal being joined.

In **pressure welding**, no filler metal is used. The union is achieved by bringing two *clean* surfaces together with sufficient pressure for a bond to be formed between them. In some pressure welding processes, a small amount of melting of the metal being joined occurs (e.g. in electric resistance welding), but in others no liquid metal is formed. The latter processes are termed **solid phase welding.** A few solid phase welding processes are carried out at room temperature, but most involve some heating of the components to be joined (but not above the melting point).

At the higher temperatures involved in fusion welding, there is greater atomic diffusion across the joined faces resulting in a more positive bond than in pressure welding, but the risk of forming brittle intermetallic compounds (especially with certain combinations of metals) is greater. The high temperatures developed often cause unwanted structural changes in the regions of the parent metal adjacent to the weld pool—called the **heat affected zone** (H.A.Z.).

Some basic forms of welded joints are shown (with associated terminology) in fig. 8.4. With increase in metal thicknesses, edge or joint preparation may be necessary in order to achieve satisfactory welds. Examples of simple edge preparation are also shown in fig. 8.4.

Fig. 8.4 Types of welded joint and edge preparation

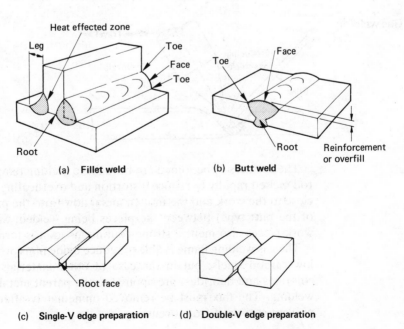

(a) **Fillet weld**   (b) **Butt weld**

(c) **Single-V edge preparation**   (d) **Double-V edge preparation**

Fig. 8.5 Fusion
welding processes

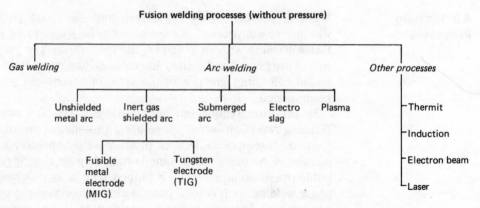

Fusion welding processes (without pressure)

Gas welding — Arc welding — Other processes

Unshielded metal arc | Inert gas shielded arc | Submerged arc | Electro slag | Plasma

Thermit
Induction
Electron beam
Laser

Fusible metal electrode (MIG) | Tungsten electrode (TIG)

8.6 Fusion Welding

Fusion Welding Processes without Pressure

Fusion welding processes may be subdivided as shown in fig. 8.5.

In **gas welding** (fig. 8.6), the required heat is produced by the controlled combustion of a mixture of oxygen and a fuel gas—usually acetylene (C_2H_2)—to give a high flame temperature (over 3000°C). The gases are fed into a torch and the relative proportions of oxygen and acetylene at the burner nozzle can be varied to give a flame that is neutral, reducing or oxidising. A *neutral flame* is the most common flame used in welding and is suitable for joining most ferrous and some non-ferrous alloys; a slightly *oxidising flame* is used for copper and its alloys in order to prevent pick-up of hydrogen; while a slightly *reducing flame* is used for aluminium and its alloys to minimise oxidation.

Fig. 8.6 Gas welding

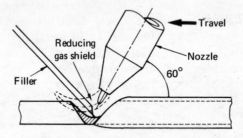

Reducing gas shield

Travel

Nozzle

60°

Filler

The surfaces to be joined are brought to melting temperature and the filler rod melted rapidly to reduce distortion and overheating. The filler rod is held close to the work and the molten metal flows into the prepared joint (usually of the butt type) between the pieces being welded. Since the edges of the work-pieces also melt, a strong continuous joint is formed.

The gas welding flame is able to reduce oxide films present on the surface of low carbon steels, but in the case of other materials reactive fluxes (e.g. chlorides and fluorides) are applied to the parent metal and filler rod before welding. The flux must be removed immediately after welding in order to prevent corrosion from occurring.

The heat input of an oxy-acetylene flame is relatively low so that gas welding is mainly used for joining thin materials and care is necessary to minimise distortion.

Gas welding is of lesser importance than hitherto, but because the equipment is inexpensive and simple to use it is still widely used in maintenance and general repair work.

Bronze welding refers to the joining of metals of high melting point (e.g. mild steel) by the use of copper-base alloy fillers. An oxy-fuel gas flame is used (e.g. oxy-butane or oxy-propane), but the process differs from true welding in that little or no fusion of the parent metal takes place.

In **arc welding**, the heat required to melt the surfaces to be joined is generated by striking an arc between an electrode and the workpiece. The high current low voltage discharge causes electrons to be transferred from the electrode via an ionised gas (known as plasma) to the workpiece, and temperatures of about 4000°C are obtainable. The electric arc may be struck between a consumable metal electrode and the workpiece, or alternatively a non-consumable electrode may be used together with a separate filler rod. Welding may be carried out manually or by the use of automatic machines. Manual arc welding is limited to relatively thin sections, while automatic machines use bigger, heavier electrodes which permit higher welding speeds so that thicker sections can be joined.

In arc welding it is important that the weld pool be protected from the ingress of gases from the atmosphere and this may be achieved either by fluxes or by the use of inert gas.

Arc Welding with Fluxes

Manual welding using an arc struck between a flux-coated metal electrode and the workpiece (usually called *manual metal arc welding*) is the most widely used fusion welding process (fig. 8.7). The metal electrode carries the current and also acts as a filler rod which deposits molten metal into the joint.

The flux coating on the electrode may be

a) Cellulose
b) Rutile (titanium oxide)
c) Acid-based (iron oxide, silicate)
d) Basic (calcium fluoride, carbonate)

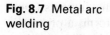

Fig. 8.7 Metal arc welding

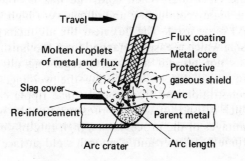

Travel

Flux coating
Metal core
Protective gaseous shield
Arc
Parent metal
Arc length
Arc crater
Re-inforcement
Slag cover
Molten droplets of metal and flux

The functions of the flux include the following:

a) To provide a protective atmosphere to exclude air (the cellulose material burns to give an atmosphere of CO_2 around the weld).
b) To cover the weld with a protective layer of fusible slag which is easily detached from the finished weld.
c) To help in stabilising the arc.
d) To act as a deoxidiser and fluxing agent for impurities in the weld pool.

Either a.c. or d.c. power supply may be used for metal arc welding, the choice depending on the metal being welded. Metal arc welding is much faster than gas welding and demands rather less skill from the operative. The main limitation of manual metal arc welding is the relatively short length of electrode which can be comfortably handled by the operative. In the execution of long welds the electrode may have to be changed several times.

Fig. 8.8 Submerged-arc welding

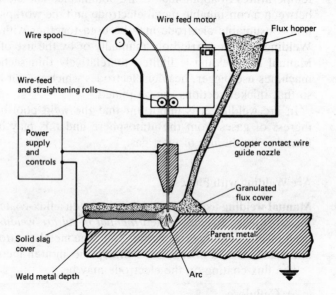

Submerged arc welding (fig. 8.8) is a mechanised version of metal arc welding. The arc is struck between a bare metal filler rod and clean parent metal under a blanket of granulated flux which is fed into the joint area just ahead of the electrode. When cold, the flux is non-conducting but, when molten, it is highly conducting and allows very high welding currents (more than 1000 A) to be used. The flux near the arc melts and forms a protective coating of slag which is easily detached from the finished weld. The very high current does not present a problem regarding ultra-violet light emission because the arc is not visible during the welding. High-quality welds of excellent penetration and free from surface ripple can be achieved by this process, which is used for welding pressure vessels, boilers and pipes.

The advantages of the process are a high rate of deposition, no visible arc with little fume or spatter and a smooth weld surface for long lengths.

Fig. 8.9 Electro-slag welding

1 Plates to be welded
2 Shoes
3 Molten slag
4 Electrode
5 Molten metal
6 Finished weld
7 Pipes for cooling
  medium

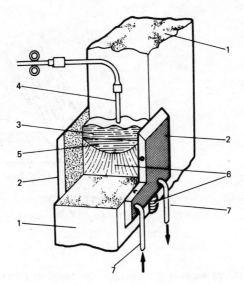

In **electro-slag welding** (fig. 8.9) the arc starts the melting process, but the slag becomes conducting at about 1000°C and the resistance of the slag generates enough heat to maintain a pool of molten metal and arcing should not then occur. The plates to be joined are placed vertically, and may be regarded as forming two sides of a "mould" into which molten weld metal is delivered in much the same way as in a continuous casting process; the other two sides of the mould are formed by water-cooled copper shoes which move vertically upwards leaving the solid weld behind. As weld metal moves slowly upwards, particles of slag and other impurities rise to the top, resulting in clean metal remaining in the weld.

As in continuous casting, the metal solidifies in a pronounced directional manner with a large grain size, so that steel welds are usually normalised in order to correct the coarse structure and thereby improve the mechanical properties.

The advantages of the method are a high rate of metal deposition (making it very suitable for thick sections) and little distortion of the fabrication.

Arc Welding using Protective Atmospheres

This group of welding processes may be subdivided into three types:

1 Inert gas (argon) shielded arc process using a consumable electrode (metal inert gas or MIG process).
2 Inert gas (argon) shielded arc process using a non-consumable (tungsten) electrode (tungsten inert gas or TIG process).
3 CO_2 gas shielded arc process using a consumable electrode.

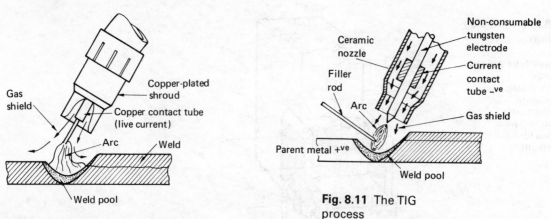

Fig. 8.10 The MIG process

Fig. 8.11 The TIG process

In the **MIG process** (fig. 8.10), a consumable electrode, in the form of a coiled, uncoated wire, is used. The installations are semi-automatic in operation and clean welds are obtained without the use of fluxes. It is a very versatile process and is widely used in heavy and light engineering.

In the **TIG process** (fig. 8.11), an external filler wire is introduced into the argon blanket either by the operative, or in semi-automatic methods through a nozzle at the front of the welding gun. For the welding of metals with tenacious films of surface oxide, instead of using the usual d.c. supply an a.c. source is used in order to remove the oxide. In order to obviate the possibility of having inclusions of metallic tungsten in the weld, a high-frequency unit is used to strike the arc. The TIG process produces high-quality welds and is sufficiently versatile to be applied to the low-current welding of foil and to high-current welding of thick sections of copper and other metals of high thermal conductivity. The process is suitable for precision work in the electronics industry.

In the **CO_2 processes** it is essential to incorporate in the composition of the filler wires deoxidising elements (e.g. Al, Si, Mn) because CO_2 can oxidise some metals (e.g. steels) at welding temperatures. The process is a modified version of the MIG process, in which argon is replaced by CO_2, and is simple and fast in operation. Semi-automatic CO_2 welding of low carbon steels is widely used and is cheaper than other arc processes.

Electron beam welding (fig. 8.12) is carried out in a vacuum, otherwise the electrons would collide with molecules of oxygen and nitrogen present in the air-space between the electron gun and the target. When a beam of fast moving electrons strikes a target, the kinetic energy of the electrons is converted into heat. The electron beam can be sharply focused on to the area to be welded so that extremely high temperatures can be produced locally. The method achieves high penetration whilst producing narrow heat affected zones and is useful for joining refractory metals (e.g. Mo, W) and for very reactive metals (e.g. Be, Zr). Electron beam welding is used for joining pre-machined articles and for the encapsulation of electronic components.

Fig. 8.12 Schematic representation of an electron beam welder

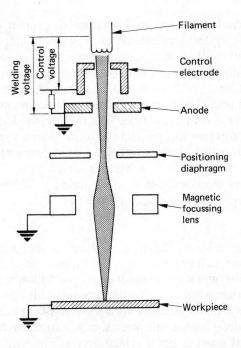

In **laser welding** (the term "laser" is an acronym for Light Amplification by Stimulated Emission of Radiation), a laser beam suitable as a heat source for welding is produced from ruby crystals in which electrons are excited. The beam of monochromatic light is focused on the area of the proposed joint. At present, laser welding is restricted to welding very thin sections including foil.

Thermit welding (fig. 8.13) relies on the heat liberated from an intensely exothermic chemical reaction, namely the reduction of iron oxide by aluminium:

$$8Al + 3Fe_3O_4 \rightarrow 9Fe + 4Al_2O_3 \qquad \begin{aligned} \Delta H &= -414.5\,\text{MJ/kmol Al} \\ &= -15.4\,\text{MJ/kg Al} \end{aligned}$$

Fig. 8.13 Thermit welding

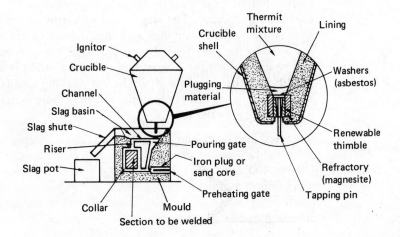

Thermit powder consists of a mixture of ground mill scale (Fe_3O_4) and aluminium powder in calculated proportions (about 3 parts Fe_3O_4 to 1 part Al). A small mould is made around the parts to be joined and above this is a crucible containing a sufficient quantity of the thermit powder. The parts to be welded are pre-heated and the thermit powder then ignited by a magnesium fuse. The heat of the reaction is so intense that the iron which is produced melts and runs down into the prepared mould, uniting with the pre-heated surfaces and producing a weld. Selected steel scrap may be added to prevent overheating of parts being joined. Some aluminium remains in the weld deposit and helps to achieve a sound joint. This method is used for welding very thick sections of steel, welding railway track on site, and for repairs to very large castings and forgings.

8.7 Pressure Welding

In pressure welding, the *clean* surfaces (either hot or cold) are pressed together and deformed enough to expose fresh, unoxidised metal. The process relies on a small amount of atomic diffusion taking place in the solid state, the joint being completed by considerable deformation. The structure of the parent metal adjacent to a solid phase weld shows evidence of cold work with significant grain directionality. In some cases a small amount of melting of the metals being joined occurs.

Fig. 8.14 Pressure welding processes

Pressure welding processes may be classified as in fig. 8.14.

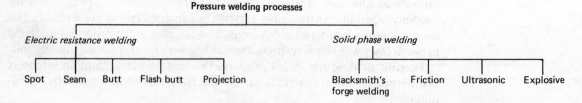

Blacksmith's forge welding is the oldest method of welding and is used for wrought iron and mild steel. The parts to be joined are heated above the recrystallation temperature and then given considerable deformation by hammering together. The slag in wrought iron acts as a flux but, for steels, sand is used which combines with iron oxide to form a fusible slag which is scattered by the hammer blows. If the oxide is not properly removed the joint will be unsatisfactory. A scarf type joint (fig. 8.15) is usually used.

Although the process is slow, the hot working results in a refined structure not obtainable by fusion methods. Further, there is much less distortion and residual stress in the joint.

Fig. 8.15 Scarf joint

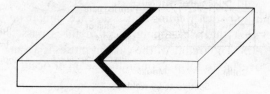

In **electrical resistance welding** methods, heat is generated by the resistance to a heavy current passing across the interface of the two components to be joined. The components are part of a low-voltage electrical circuit, the high current being supplied to the electrodes through a transformer. The areas to be welded are cleaned to remove surface impurity, then pressed together before a current is passed; the pressure is maintained during the current flow and for a very short time afterwards. The heat developed during the passage of the current is enough to melt the metal in the area of the electrodes and thus a good strong weld is produced. The electrodes vary in shape, and must have adequate electrical and thermal conductivity as well as mechanical strength and wear resistance. Several copper alloys have been developed for electrodes, notably chromium copper (Cu–0.5% Cr) and sintered tungsten-copper compacts. Low carbon steels are easy to weld by these methods, but high conductivity metals (e.g. Cu, Al) are more difficult. Spot, seam, stitch and projection welding are commonly used resistance welding techniques for joining sheet metal.

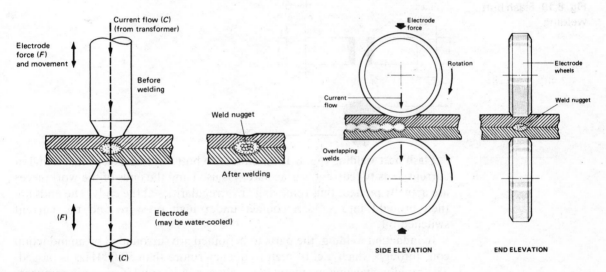

Fig. 8.16 Spot welding

Fig. 8.17 Seam welding

In **spot welding** (fig. 8.16), the sheet metal surfaces overlap and are clamped between a pair of water-cooled electrodes. After each weld the material or the welding gun is moved to the next position. The time taken for each weld is usually less than 0.5 sec, and machines which can perform a very large number of welds simultaneously are used in the production of car-bodies and hot-water radiators.

Seam and stich welding (fig. 8.17) are similar in principle to spot welding except that the overlapping sheet metal is passed between rotating wheel electrodes. The timing of the current pulses is arranged so that the spot welds either overlap to form a continuous seam or occur at regular intervals to form a stitch.

Fig. 8.18 Projection welding

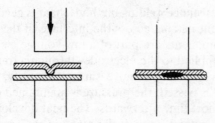

In **projection welding** (fig. 8.18), small projections are embossed on one of the sheet components. The current flow, and hence the resulting heat is localised in these projections, which collapse under pressure to give a sound joint.

Butt welding is used to join lengths of rod, tube or wire. The ends are pressed together and then current passed through the material so that the ends are heated to a plastic state due to the higher electrical resistance existing at the area of contact. The pressure is sufficient to form a weld.

Fig. 8.19 Flash butt welding

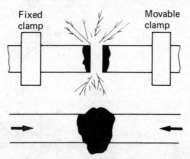

Fixed clamp · Movable clamp

Flash butt welding (fig. 8.19) is similar to butt welding except that a higher current density (current/surface area) is used and the ends of the workpieces are actually melted, thus removing any irregularities at the ends. The ends are then brought into sudden contact under high pressure and the current switched off.

In **induction welding**, the parts to be joined are surrounded by an induction coil, through which a.c. of high frequency (more than 50 000 Hz) is passed. The rapidly changing magnetic field causes a very rapid increase in temperature up to that required for welding, after which pressure may be exerted to complete the joint. For some purposes induction heating is more convenient than using an arc or a flame.

Cold pressure welding is suitable for producing small joints in aluminium, copper and mild steel components and is used in the production of longitudinal welds in aluminium cable sheathing, butt welds in copper conductors, and for spot pressure welding of sheet metal. Thorough cleaning of the surfaces is essential and may involve acid pickling to remove oxides, solvent degreasing to remove oil and grease, followed by scratch-brushing. Considerable pressure is required to complete the joint and this may be applied through punches and dies. The pressure causes severe distortion which must break up any remaining surface oxide films to ensure a satisfactory bond. Deformation in the range 50–95% is generally required.

Fig. 8.20 Friction welding

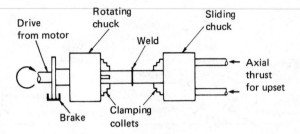

In **friction welding** (fig. 8.20), one of the pieces to be joined is rapidly rotated against the other piece, which is held in a fixed chuck and pressed firmly against the rotating piece. The frictional heat generated causes softening of the metal and the break-up of oxide films. As soon as a sufficiently high temperature is reached, the rotation is stopped and pressure applied. A small amount of atomic diffusion takes place across the clean surfaces of the metal and a good sound joint is obtained. Friction welding is useful in joining dissimilar metals, e.g. copper and aluminium.

Fig. 8.21 Explosive welding

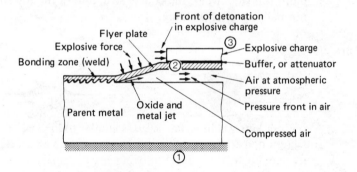

In **explosive welding** (fig. 8.21), a high-energy shock wave cleans the surface and impact achieves a satisfactory joint despite the fact that there is little heat generated. Welding occurs as a result of high-velocity oblique impact of two plates of metal. The method is used for the cladding of one metal plate on to another.

In **ultrasonic welding**, induced mechanical vibrations break up surface oxide films and heat up the metal. The joint is completed by the application of pressure. The method is suitable for joining metal foil.

8.8 Structure of Weld Metal and the Heat Affected Zone

In a welded joint, a variety of metallic structures may be represented including examples of cast, wrought and heat-treated conditions. In a fusion weld, the weld metal itself will have a typical cast structure with its inherent defects. The regions of the parent metal which have undergone unwanted structural changes due to high temperatures are called the heat-affected zone (H.A.Z.), while the unaffected part may show a typical wrought structure.

Figure 8.22 shows the principal features of fusion welds.

Fig. 8.22 Main features of fusion welds

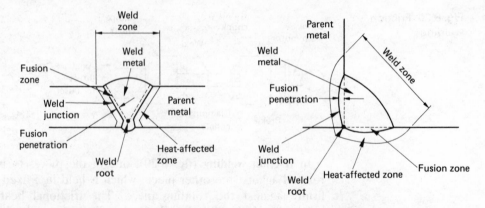

Weld metal   The weld metal is, in effect, a miniature casting which has cooled rapidly from a very high temperature. Long columnar crystals may be formed giving a rather weak structure (fig. 8.23a). In a steel weld made with more than one run, each deposit normalises the preceding run (fig. 8.23b), resulting in grain refinement and improved mechanical properties. In a multi-run steel weld, the coarse structure of the top run may be corrected by normalising after welding.

Fig. 8.23 Typical arc weld structures

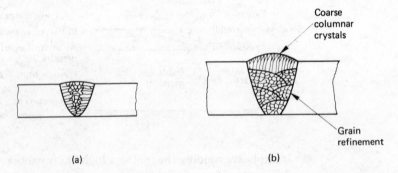

(a)   (b)

The possible defects in a weld metal include the following: poor structure, nonmetallic inclusions, gas porosity and cracks.

Inclusions of oxides and nitrides must be avoided by protecting the weld by the use of fluxes or inert gases. Slag inclusions can be avoided in multi-run welds by removing each layer of slag after each deposit.

Gas porosity is mainly caused by the presence of hydrogen in the weld metal or the reaction between hydrogen and the oxide present in the melted parent metal to form steam. There are several possible sources of hydrogen during welding, including the flame in gas welding, dampness of materials used, and the flux coating in metal arc welding. Hydrogen is readily soluble in the liquid state, but only slightly soluble in the solid state. Some of the hydrogen evolved during solidification may be trapped in the solid metal resulting in gas porosity in the weld. However, in the welding of tough pitch copper, which contains about 0.04% oxygen present as Cu_2O, the hydrogen may react with the Cu_2O to form steam, resulting in unsoundness.

Cracking may also occur in the weld metal, especially in joints prepared under constraint, due to the contractional stresses producing during cooling. The tendency towards this hot cracking is greatly influenced by the grain size and the presence of impurities at the grain boundaries. A small-grained material with a large amount of grain boundry area is more able to accommodate strains than a coarse-grained material. The presence of low melting point or brittle impurities at the grain boundaries is conducive to cracking. Hence judicious choice of welding parameters, careful selection of filler rod metal containing suitable deoxidants, and protection of the molten weld pool by fluxes or suitable inert gas do much to eliminate weld defects.

The heat-affected zone of the parent metal   The extent of structural changes in the parent metal will depend on the temperature attained and the time at that temperature, so that the following factors are important: method of welding, the possible use of heating prior to, or after, welding (preheating or post-heat-treatment), speed of welding, dimensions of material being joined and its properties, including thermal conductivity and specific heat.

Welding joints prepared from metals of high thermal conductivity (e.g. Cu, Al) have wider heat-affected zones than those prepared from steel. An increase in the speed of welding reduces the width of the H.A.Z., as does an increase in the intensity of the heat source. Hence, other factors being equal, an arc weld produces a narrower H.A.Z. than gas welding.

As far as the effect of weld heat on properties is concerned, parent metals may be divided into three groups:

1 Solid solutions.
2 Age-hardenable alloys.
3 Alloys which undergo a martensite type of transformation on cooling (e.g. plain carbon steels).

Solid solutions are often used in the cold worked condition in order to obtain the required strength. The effect of weld heat is to cause recrystallisation and grain growth resulting in softening of the material (fig. 8.24). With the consequent increase in ductility it is unlikely that any cracking will occur in the H.A.Z.

Fig. 8.24 Effects of welding on the structure and hardness of cold worked pure metal

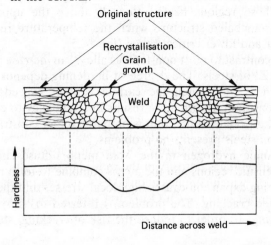

Fig. 8.25 Effects of welding on the structure and hardness of normalised mild steel

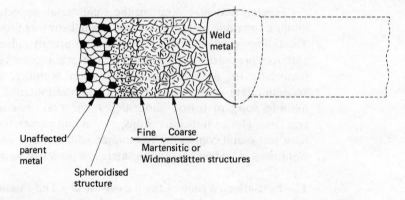

Unaffected parent metal

Spheroidised structure

Fine   Coarse
Martensitic or Widmanstätten structures

Weld metal

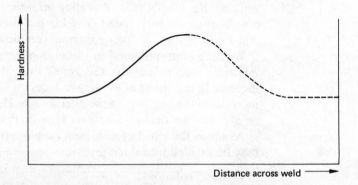

Distance across weld ⟶

Age-hardenable alloys may be seriously affected in the H.A.Z. In addition to softening and grain growth, some incipient fusion of intermetallic compounds near the fusion line and overaging further from the weld metal may occur. These effects constitute likely causes of failure. Hence some aluminium age-hardening alloys (e.g. Duralumin) are not recommended for fusion welding, but others are perfectly suitable.

Steels form the most important group of weldable alloys and the H.A.Z. will show a variety of structures, ranging from coarse, overheated structure for those regions heated to well above the upper critical range to an under-annealed structure when the temperature reached lies between the upper and lower critical range (fig. 8.25).

In contrast to most non-ferrous alloys, an *increase* in hardness occurs in the H.A.Z. of steels. The degree of hardening depends on the carbon content, and it is only when the carbon content exceeds about 0.3% that the martensitic transformation results in a tendency to cracking in the hardened zone, especially if the joint is prepared under constraint. The welding of low carbon steels presents no problems.

Atomic hydrogen in the weld metal diffuses into micro-cracks in the martensitic regions and the atoms combine to form molecular hydrogen; the resulting expansion causes high local stresses and these regions are prone to delayed cracking. The problem is referred to as *hydrogen-induced H.A.Z. cracking*. In order to reduce the risk of cracking, steels are heated to about

200°C immediately before welding and may be re-heated to 200°C after welding to relieve residual stresses. Also, slow welding speeds and the use of electrodes with low hydrogen coatings or austenitic electrodes are effective in reducing the cracking tendency. Hydrogen is soluble in austenite (but not in martensite or ferrite) so that the use of austenitic electrodes allows the retention of most of the hydrogen in the weld metal.

8.9 Weld Defects

Standard terminology and definitions of a wide range of faults are given in BS 499: Part I. The International Institute of Welding classifies faults as follows:

1 Cracks
2 Cavities
3 Solid inclusions
4 Lack of fusion and penetration
5 Imperfect shape
6 Miscellaneous faults.

In practice gross defects may be removed by grinding and the weld then repaired.

Cracks

Surface cracks are detected by visual examination, magnetic particle or penetrant inspection, whereas the detection of internal cracks requires ultrasonic or radiographic techniques. Examples within this group include solidification cracking, hydrogen-induced H.A.Z. cracking (see section 8.6), and lamellar tearing.

Fig. 8.26 Solidification cracking

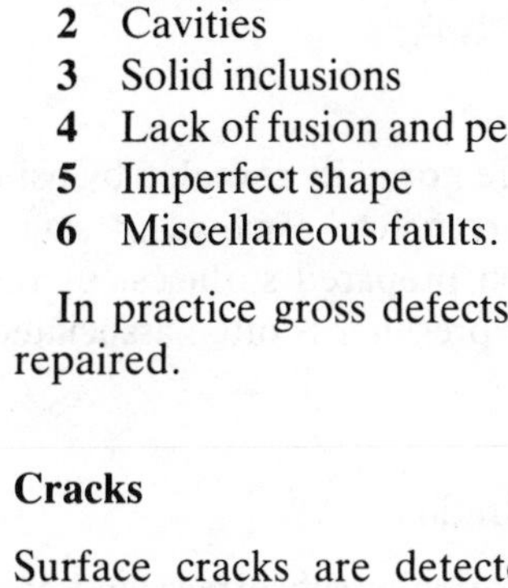

Fig. 8.27 Lamellar tearing

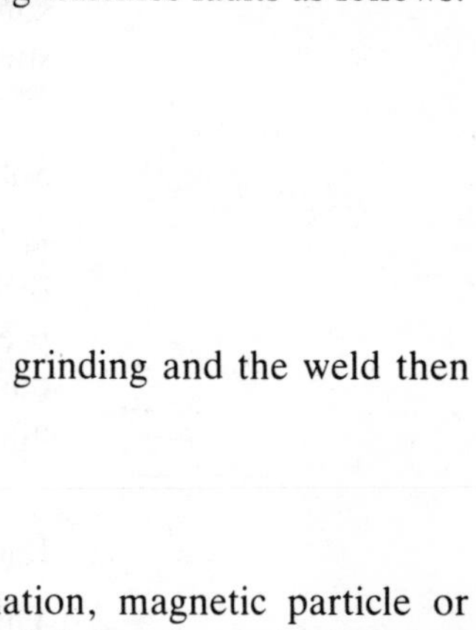

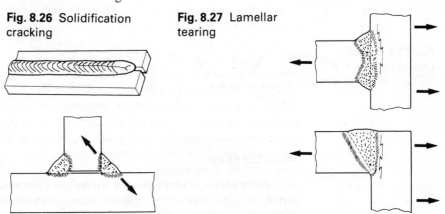

Solidification cracking which occurs in the weld metal (fig. 8.26) is caused by a large depth/width ratio of weld bead, high arc energy and/or preheat, or the pick-up of sulphur, phosphorus or niobium from the parent metal. *Lamellar tearing* (fig. 8.27) is caused by poor ductility in the through-thickness direction of the plate due to non-metallic inclusions. The crack path is generally step-like in character.

Cavities

Surface cavities are detected visually and internal cavities by using ultrasonics or radiography. The types of cavity occurring in weld metal include uniformly distributed porosity, worm holes, and surface porosity.

Uniformly distributed porosity results from the entrapment of gas in solidified weld metal. The gas may originate from dampness or grease on the workpiece or consumables, or nitrogen pick-up from the atmosphere. It is possible for carbon monoxide to cause porosity if insufficient deoxidant is used in the filler wire. *Worm holes* result from the entrapment of gas between solidifying dendrites of weld metal. *Surface porosity* may be due to excessive surface contamination or poor control at the weld start or end.

Solid Inclusions

Solid inclusions of slag are normally revealed by using radiography. Such slag inclusions are due to incomplete removal of slag in multi-pass welds, the presence of scale/rust on prepared surfaces, or the use of electrodes with damaged coatings. Their presence is often associated with imperfect shape in underlying passes.

Lack of Fusion and Penetration

This type of defect is normally subsurface and therefore only detected by ultrasonic or radiographic methods. The problems are caused by incorrect welding conditions (e.g. insufficient heat input) or incorrect weld preparation (e.g. root face too large). Figure 8.28 illustrates typical examples.

Fig. 8.28 Typical examples of lack of fusion and penetration

a) Lack of side wall
    fusion

b) Lack of inter-run fusion

c) Lack of penetration

Imperfect Shape

All shape defects are detected visually. Typical examples include linear misalignment, excessive reinforcement, overlap, undercut and excessive penetration, which are illustrated in fig. 8.29.

Miscellaneous Faults

Such faults include accidental *arc strikes* and *spatter* of globules of molten metal which adhere to the parent metal remote from the weld. Again problems of *distortion* may occur if the heating or cooling is uneven.

Fig. 8.29 Examples of imperfect shape

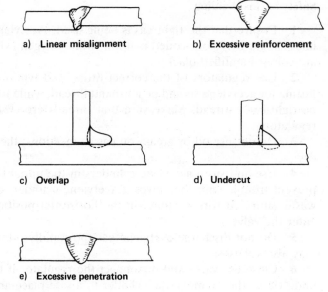

a)   **Linear misalignment**

b)   **Excessive reinforcement**

c)   **Overlap**

d)   **Undercut**

e)   **Excessive penetration**

8.10 Safety Precautions

The precautions to be observed include general matters as well as requirements specific to a certain joining process. Because of the much higher temperatures involved, the hazards are greater in welding than in some other processes.

General Safety

Typical safety instructions issued to an operative should include the following:

1   Be familiar with the positions of all control and emergency switches, exits, fire-fighting appliances and first-aid points.

2   Wear appropriate protective clothing.

3   Do not carry out any joining process (especially welding) in a confined space.

4   Be mindful that toxic vapours or gases may be evolved.

5   Hot metal should be suitably marked and set aside. Do not touch any metal before checking (e.g. with a crayon) that it is cold.

6   Do not leave flammable material in an area where joining processes are being carried out.

7   Clean any article that has been in contact with flammable material before joining.

8   Comply with any instructions issued regarding the work area, work method and equipment.

Safety in Gas Welding

1 Ensure that the right gas is being used. Acetylene cylinders and pipe lines should be colour-coded maroon, while oxygen cylinders and pipe lines are colour identified black.

2 Use regulators of the correct fitting and type for the particular gas. Fitting for acetylene should be left-hand thread, while for oxygen they should be right-hand thread. Make sure that threads are clean before connecting regulators.

3 Do not use oil or grease on gas connections; they must be assembled dry.

4 Use and store acetylene cylinders in the upright position, in order to prevent trouble with the valves. (Acetylene cylinders are filled with cotton waste saturated with acetone; in the horizontal position cotton waste may enter the valve.)

5 Do not discharge acetylene gas too rapidly, otherwise some acetone may also volatilise.

6 Close the valve and disconnect the regulator if the cylinder becomes unduly hot, then remove the cylinder to a safe place and allow to cool.

7 Do not use oxygen as a substitute for compressed air. (Oxygen enrichment will cause clothing to burn explosively.)

8 Keep gas hoses clear of working areas as far as possible to avoid tripping hazards and abrasions of their surface.

9 Light the acetylene, then open the oxygen valve until the correct flame is obtained. "Flash-back" arrestors should be fitted to the acetylene supply line.

10 Take care in handling the lit torch.

11 Wear goggles and appropriate protective clothing.

Safety in Arc Welding

1 Check that the welding bench is earthed by a separate earth cable to avoid hazards from equipment faults. Clothes, shoes and gloves must be kept dry; they will then give extra insulation.

2 Do not look directly at an arc—use a face-shield with a suitable filter (conforming to BS 679), and have all exposed skin adequately covered. Arcs emit ultra-violet and infra-red radiation which can cause skin-burns and eye irritation ("arc-eye").

3 Check that the work is stable before striking an arc.

4 Wear appropriate protective clothing. In arc welding there is a considerable amount of spattering of drops of molten metal so that gauntlets, aprons, spats and safety shoes are advisable.

5 Ensure that fume extraction equipment is working properly when using flux-shielded processes, inert gases, or welding materials which emit harmful fumes.

6 Keep electric cables clear of working areas.

Fig. 8.30 Typical machining processes

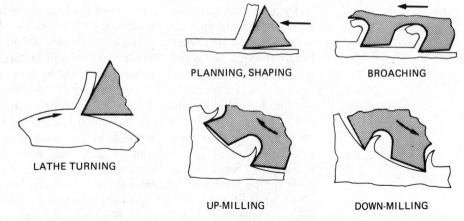

PLANNING, SHAPING   BROACHING

LATHE TURNING

UP-MILLING   DOWN-MILLING

8.11 The Machinability of Metallic Materials

Traditional machining is a cold working process in which a wedge-shaped cutting tool is forced into the stock to remove thin layers of materials as broken chips or as a continuous ribbon of swarf. The removal of material in chip form is much preferred. The machining processes are secondary operations applicable to the many products where close dimensional accuracy is necessary. In such processes either the workpiece or tool rotates or reciprocates in order to bring about the cutting action. Figure 8.30 illustrates some basic processes.

Modern production practice makes severe demands on the cutting tool. The material used for such tools should possess a low coefficient of friction, good abrasion resistance, the ability to resist softening at high temperature, and sufficient toughness to resist fracture. The principal cutting tool materials include:

1 *High Carbon Steels* (0.8–1.2% carbon). Because of its low hardenability such material can only be used for small tools. Again, softening occurs at around 300°C so that the material is unsuitable for high-speed and heavy-duty work.

2 *High Speed Steels.* Perhaps the best-known material in this group is the 18-4-1 high speed steel (18% W, 4% Cr and 1% V). Other materials in this group may contain significant percentages of molybdenum and/or cobalt.

3 *Carbides.* Carbides usually bonded with cobalt are formed into tool tips which are brazed to a steel base. For example, 82% tungsten carbide with 10% titanium carbide bonded with 8% cobalt is suitable for machining steel.

In similar fashion diamonds or certain ceramics, e.g. Al_2O_3, may be mixed with a binder and processed into a cutting tool.

The ease with which a material may be machined and the type of surface finish obtained depend on many factors, which may be classified as mechanical engineering factors and metallurgical factors. The engineering factors include the design of the cutting tool, the speed and depth of cut, and the method of lubrication, but a discussion of these are not within the scope of this book.

The microstructure and properties of the material being machined also play a very important part and an account of these factors is given below.

In metal cutting, the material immediately ahead of the tool becomes highly stressed and a soft, ductile material will undergo considerable plastic deformation so that a continuous ribbon will be formed. A very soft, highly ductile material spreads under the pressure of the cutting tool, which tends to become buried in the material, so that a tearing action instead of a cutting action takes place. In the case of a much less ductile material, considerable work hardening will take place and small discontinuous chips will be broken off. This type of material can be machined much more quickly, and is required for work on automatic machines. For easy machining, a material must not be too hard, otherwise there is excessive wear on the cutting tool. The best combination of mechanical properties for good machinability is low hardness with fairly low ductility. For example, magnesium alloys have these characteristics and possess excellent machinability.

The machinability of soft materials can be improved by introducing local brittleness into the alloy without causing significant reduction of the toughness of the material as a whole. This can be achieved in the following ways:

1   By arranging to have a separate constituent in the microstructure
Small isolated particles of the separate constituent act as stress raisers and minute fractures travel from the cutting edge of the tool to these particles, thus decreasing the friction between the tool and the material being machined. This will reduce the power required for cutting and also the wear on the tool. Because the continuity of the structure is broken by these particles, the swarf produced will be in the form of small chips.

Examples of materials to which deliberate additions are made in order to form a separate constituent, and thus improve the machinability, are as follows:

About 0.5–3.0% lead may be added to copper-base alloys; the lead is insoluble in both the liquid and solid state and is present in the structure as rounded particles. About 1% tellurium is added to copper used for electrical work. Unlike lead, tellurium improves the machinability without significantly decreasing the electrical conductivity.

About 0.2–0.6% sulphur is added to low carbon steels together with about 1% manganese in order to ensure that the sulphur is present as the soft, plastic MnS and not the objectionable, brittle, low melting point FeS. An addition of 0.15–0.30% lead may be added to certain alloy steels as a preferred alternative to increasing the sulphur content, which can adversely affect toughness. Alloys containing several separate constituents usually machine easily because the continuity of the structure is broken up. For example, the presence of graphite in cast iron, or pearlite in steel, and of $CuAl_2$ in Duralumin all contribute to the good machinability of the respective materials.

2   By cold working before machining
Products such as bars and rods (in relatively small sizes) intended for machining operations are often supplied in the cold drawn ("bright drawn")

condition. The cold drawing process improves the machinability by reducing the ductility of the surface layers of the product. Moreover, the cold working results in products with closely controlled dimensions.

Sulphur-bearing free cutting steels are usually supplied in the bright drawn condition.

3 *By heat-treating before machining*

Low carbon steels are more easily machined in the normalised condition, i.e. when the grain size is small and the pearlite areas are small and uniformly distributed.

High carbon steels are best given a spheroidising annealing treatment before machining [see page 109 of *Structure*]. The treatment consists of heating the steel for a very long time (e.g. 24 hours) at a temperature just below the Ac_1 point (e.g. 680°C) to convert the cementite into rounded globular particles. In the spheroidised condition, the steel has very good machinability, but the toughness is poor. Therefore, after machining, the steel must be given a corrective heat treatment (e.g. hardening and tempering) to restore the mechanical properties.

In general, soft ductile non-ferrous materials are more easily machined when the grain size is small. However, the machinability of the harder alloys is often best when the grain size is large.

Exercises 8

1   List four types of joining processes.

2   Make sketches of four types of rivet.

3   *a*) Name two types of adhesive suitable for joining metals.
     b) State one advantage of adhesive joining.

4   Make sketches of four types of joint used in soldering.

5   Name four different methods of applying solder to a joint.

6   *a*) State the composition of (i) Tinman's solder, (ii) Plumber's solder.
     b) Make a drawing of the lead-tin thermal equilibrium diagram.
     c) Draw cooling curves from the liquid to the solid state of each of the alloys in *a*).
     d) Name one common application of each of the alloys in *a*).

7   *a*) Name two types of brazing alloy.
     b) State an approximate range of temperature in which each of the alloys in *a*) melt.
     c) Name two fluxes commonly used in brazing.

8   Name six different sources of energy in welding.

9   List three types of flame used in gas welding and name one material commonly welded by each flame.

10   *a*) Name four types of flux coating which may be present on a welding electrode.
     b) Name four functions of the flux coating.

11   With the aid of sketches distinguish between the MIG and TIG welding processes.

12   *a*) Explain what is meant by "solid state welding".
     b) Give four examples of solid state welding processes.
     c) Briefly describe one process mentioned in *b*).

13   *a*) Name four characteristics required of a material suitable for use as a cutting tool.
     b) Name two types of tool material.

14   *a*) Name two ways in which the machinability of a highly ductile material may be improved.
     b) Give one example of each of the methods mentioned in *a*).

15   In order to achieve easy machinability state the condition in which the following materials should exist:
     a) low carbon steel, *b*) high carbon steel.

9 Corrosion of Metals

Corrosion may be defined as the reaction of a metal with its environment and which results in the deterioration of the properties of the metal. The reaction results in the formation of a chemical compound, known as a corrosion product. It has been estimated that corrosion costs the U.K. alone about £4000 million a year (1980) in metal losses and protective measures, and that about 25% of this cost could be saved by proper application of known prevention methods.

Corrosion reactions may be divided into two main groups:

1   Dry corrosion where the metal reacts with its gaseous environment (e.g. high temperature oxidation).
2   Wet corrosion (electrochemical corrosion) where the metal reacts with an aqueous environment.

Although oxidation in the theoretical sense of removing electrons to convert atoms into positively charged ions occurs in all forms of metallic corrosion, it is usual in metallurgy to restrict the term oxidation to chemical reaction in the absence of moisture.

9.1 Oxidation (Dry Corrosion)

Oxidation implies a reaction between a metal and an atmosphere which may include one or more of the following gases: oxygen, carbon dioxide, sulphur dioxide. Reaction may take place at ambient or elevated temperatures. Oxidation is a special case of corrosion where electrochemical processes take place within the corrosion product (e.g. an oxide or sulphide layer).

Most metals possess an affinity for oxygen and tend to react with air or oxygen to produce a surface film of oxide, e.g.

$$x\mathrm{M} + \frac{y}{2}\mathrm{O}_2 \rightarrow \mathrm{M}_x\mathrm{O}_y$$

An oxide layer which is more than 10^{-3} mm thick is usually referred to as a "scale". The term "scale" often denotes a thick, visible oxide layer formed under highly oxidising conditions. Tarnishing usually refers to a discoloration or dulling of a metal surface due to the presence of an oxide film. With most metals the amount of oxidation which takes place at ordinary temperatures is not serious. In fact, in many cases, the oxide layer which rapidly forms on a freshly exposed metal surface tends to protect the metal from further oxidation.

In the case of iron heated in air, the *anode* reaction is

$$\mathrm{Fe}_{\mathrm{solid}} - 2e \rightarrow \mathrm{Fe}^{2+}_{\mathrm{scale}}$$

Fig. 9.1 Electrode
positions in oxidation

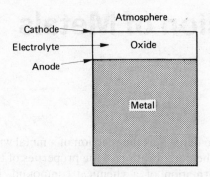

which takes place at the metal-scale interface.

The *cathode* reaction is

$$\tfrac{1}{2}O_{2\,gas} + 2e \rightarrow O^{2-}_{scale}$$

The scale itself acts as the electrolyte and the electron carrier as well as forming a barrier between the metal and atmosphere (fig. 9.1). In order for oxidation to continue, either oxygen must diffuse through the layer to the metal surface, or metal ions must diffuse through to the surface of the oxide layer. In many cases it is diffusion of metallic ions which occurs to the greater extent owing to their relatively small size. In this case the oxidation rate will decrease as the thickness of the oxide layer increases; at a given temperature the oxide thickness varies as $\sqrt{t}$ where t is the time, thus following a parabolic law. In other cases, however, the oxide layer may be porous or it may not adhere to the metal surface. In these circumstances there will be free access of oxygen to the metal surface and the thickness of oxide formed at a given temperature will vary linearly with time. It is seen, therefore, that the properties of the oxide layer formed are of great importance and control the rate of oxidation in a particular environment. In illustration of the above, consider the scaling of iron. In this case the oxide layer is loose and porous, so that oxidation will continue and the scale will thicken until the complete cross-section of metal will be oxidised. In the case of aluminium, however, the oxide layer is adherent and non-porous, so that the thin film first formed will act as a protection to the underlying metal.

The rate of oxidation of metals increases rapidly as the temperature is increased. Within certain alloys, selective oxidation may occur, e.g. pearlite zones in steel are more readily oxidised than other constituents, thus resulting in the decarburisation of surface regions in an oxidising atmosphere.

Oxidation rate is usually measured by weighing specimens periodically after exposure to a given temperature to determine the amount of oxide formed.

In order to control the amount of oxidation it is necessary for the oxide film formed to be protective. For example, although aluminium has great affinity for oxygen, the oxide formed is a very good electrical insulator, i.e. it is highly resistant to the passage of electrons and ions, thus giving a highly protective film. It follows that an important method of achieving oxidation resistance is by introducing appropriate alloying elements in order to produce protective films, e.g. the addition of about 1% aluminium to copper-base alloys. Larger additions are made to iron and steel to provide significant oxidation resistance. Although Al and Si form protective films, in the amounts required they embrittle the metal and this restricts their use. The most common addition to iron and steel is chromium, and about 12% is added in stainless cutlery and up to about 20% in heat-resisting steels.

When ordinary iron is heated at temperatures above about 600°C a thick scale is formed, consisting of FeO next to the metal, Fe_2O_3 on the outside, and Fe_3O_4 in an intermediate layer. When the above amounts of chromium are added, the inner layers are replaced by a layer of protective Cr_2O_3, beneath a thin outer layer of Fe_2O_3. The outstanding protective qualities of Cr_2O_3 at high temperatures make chromium an essential alloying addition for most heat-resistant alloys, such as nichrome electric heater wire (80% Ni, 20% Cr) and various alloys for gas turbines [see pages 166–67 *Structure*].

9.2 Electrochemical Corrosion

Aqueous corrosion is a complex problem because of the many variables involved, but the following parameters are basic to the problem:

a) The material used.
b) The design and fabrication of the manufactured product.
c) The working environment.

Experimental work shows that the actual attack on a metal is electrochemical in character. The presence of moisture is a vital factor, and oxygen also plays a very important part in the process. For example, iron will not rust in dry air, nor in pure water alone, but when air and moisture are present together iron rusts very quickly. Rusting continues unabated because the layer of corrosion product formed is loose and porous.

In the presence of water all metals have a tendency to dissolve or corrode, during which the metal discharges positively-charged ions into solution. This leaves the metal with a negative charge. The greater this negative charge the greater is the tendency of the metal to dissolve or corrode. For example, if a piece of zinc is immersed in an aqueous solution of zinc sulphate containing a definite concentration of ions, it is found that there is only one electric potential at which equilibrium can exist between the metal and the solution. This is because the zinc, by tending to dissolve, forms zinc ions until an equilibrium condition is reached between the liquid and the metal. The magnitude of the resulting negative charge on the metal is characteristic of the particular metal and is called its electrode potential. If a suitable scale is chosen this characteristic electrode potential can be expressed in volts, as may be done by connecting the zinc electrode to a "hydrogen electrode" whose potential is taken as zero. The hydrogen electrode consists of platinum foil coated with finely divided platinum (to give it a large surface area) suspended in a molar solution of hydrogen ions. In this way metals can be arranged in the order shown in table 9.1 according to their electrode potentials, giving a list known as the **electrochemical series**.

Table 9.1 Electrochemical Series

			Metal	Electrode potential (volts)
Most electropositive	Reactive metals	Anodic end (corroded)	Sodium	−2.71
			Magnesium	−2.40
			Aluminium	−1.70
			Titanium	−1.63
	increasing reactivity		Zinc	−0.76
			Chromium	−0.56
			Iron	−0.44
			Cadmium	−0.40
			Nickel	−0.25
			Tin	−0.14
			Lead	−0.13
			Hydrogen	0.00
	increasing nobility		Copper	+0.35
			Silver	+0.80
			Platinum	+1.20
			Gold	+1.50
Least electropositive	Noble metals	Cathodic end (protected)		

9.3 Electrochemical Cells

In aqueous corrosion, chemical changes occur which induce the flow of an electric current. In order to produce a continuous flow of current, a complete **electrochemical cell** must be set up, i.e. there must be an anode, a cathode and an electrolyte (fig. 9.2). The anode corrodes whereas the cathode is not consumed. Such an electrochemical cell produces an electric potential from a chemical reaction. In contrast, an **electrolytic cell** is one in which a chemical reaction is caused by an externally applied electrical voltage so that current is consumed.

As stated previously the discharge of positively charged ions into solution leaves the metal negatively charged, so that the *anode* in an electrochemical

Fig. 9.2 Comparison of electrolytic and electrochemical cells

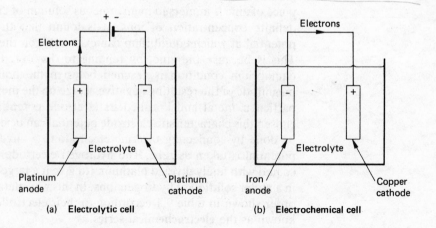

(a) **Electrolytic cell**

(b) **Electrochemical cell**

cell has *negative* polarity. (In an electrolytic cell, electrons produced at the anode are "pumped away" by the externally applied current, thus making the anode *positive*.)

It is clear that the driving force of electrochemical corrosion is the potential difference between the anode and the cathode, and the current flowing in the circuit determines the corrosion rate. If the circuit resistance is increased, or the potential decreased, the current is reduced and therefore the corrosion rate is reduced. The corrosion rate is also affected by the relative areas of anode and cathode. In general, for a given area of anode, the attack increases as the area of the adjacent cathode is increased.

Consider now the various reactions which can take place at the electrodes of an electrochemical cell. At the *anode*, electrons are produced by such means as:

(i) Metal may go into solution, e.g. $M \rightarrow M^{n+} + ne$
(ii) Anions may be attracted to the anode, resulting in the formation of an oxide, hydroxide or an insoluble salt e.g.

$$2M + 2(OH)^- \rightarrow M_2O + H_2O + 2e$$
$$M + (OH)^- \rightarrow MOH + e$$
$$M + X^- \rightarrow MX + e$$

At the *cathode*, electrons are consumed and this may be achieved in several ways depending on the environment; the ways include the following:

(i) In acid solutions, hydrogen may be evolved and oxygen may be reduced:

$$2H^+ + 2e \rightarrow H_2$$
$$O_2 + 4H^+ + 4e \rightarrow 2H_2O$$

(ii) Oxygen reduction may also occur in netural or alkaline solutions:

$$O_2 + 2H_2O + 4e \rightarrow 4OH^-$$

(iii) Metal deposition: $M^{n+} + ne \rightarrow M$
(iv) Metal ion reduction, e.g. $M^{3+} + e \rightarrow M^{2+}$.

The ability of metals to resist corrosion depends to some extent on their position in the electrochemical series. The farther two metals are separated from one another in this series, the more powerful is the electric current produced by their contact in the presence of an electrolyte, and hence the more severe the corrosion. Also the more rapidly the baser metal (the anode) is attacked, the more will the nobler metal (the cathode) be protected. However, it is not always possible to predict from the electrochemical series which metals will be anodes or cathodes when joined or what the cell voltage will be. For example, if a highly reactive metal is covered by a protective oxide film, its ions cannot easily leave it and go into solution even though the metal is readily ionizable. Thus oxide-coated aluminium acts as the *cathode* to a zinc *anode*, despite the positions of these metals in the electrochemical series. Also 18/8 stainless steel usually exists in the noble (passive) state, but if its oxide film is destroyed it becomes reactive and much more susceptible to attack.

By taking pairs of metals and alloys in a given medium such as sea-water and observing which becomes the anodes, a galvanic series can be prepared, similar to, but not identical with, the electrochemical series. Table 9.2 lists materials in order of decreasing reactivity. For example, brass fittings in contact with steel increase the corrosion of the latter in sea water.

Table 9.2 Galvanic series in sea water (pH 8.1–8.3)

Anodic end	Magnesium
	Zinc
	Aluminium
	Plain carbon steel
	Cast iron
Decreasing	Active 18/8 stainless steel
reactivity	Brass
	Copper
	Aluminium bronze
	Cupro-nickel
	Nickel
	Silver
	Passive 18/8 stainless steel
	Monel metal
Cathodic end	Titanium

Passivity

Strong oxidising agents can oxidise the surface of some metals and alloys (e.g. Al, Fe, stainless steels) and a tenacious oxide film may be formed. The metal then becomes immune to further attack and behaves like a noble metal. The metal is said to be "passivated". For example, pure aluminium is particularly resistant to concentrated nitric acid. The latter acid also passivates iron by forming a film of oxide on its surface. However, the film of iron oxide is brittle and fragile, and also does not give much protection against corrosion by non-oxidising acids such as HCl or H_2SO_4.

The ability of a passive metal to withstand corrosion depends to a large extent on the properties of the film including the following: thickness, adhesion to basis metal, mechanical strength, and the self-healing qualities.

9.4 Other Types of Corrosion Cell

The formation of an anode and cathode need not necessarily be due to contact between dissimilar metals. An electrochemical cell may be set up between two portions of the same piece of metal if there is some difference between the portions. This difference may be one or more of a large number of factors including the following:

a) The presence of particles of another phase or the existence of segregation. In a two-phase alloy, or a metal containing impurities, one phase may be anodic with respect to the other and electrochemical micro-cells may be established. For example, in the pearlitic areas in a steel, the ferrite is anodic with respect to cementite, while during the cooling of austenitic stainless steel welds deposition of chromium carbide may occur at grain boundaries in the heat-affected zones. This deposition depletes the chromium content in the immediate vicinity, causing it to become anodic relative to the chromium-rich austenite. In a corrosive environment therefore, the grain boundary regions are attacked. Corrosion of the latter type is referred to as "weld decay". Usually, therefore, pure metals or single-phase alloys tend to have better corrosion resistance than impure metals or multi-phase alloys.

b) Randomly arranged grain boundary atoms tend to ionise more rapidly than atoms within a regular crystal lattice, so that a grain boundary tends to be anodic with respect to the grain itself. Also, grain boundaries are sites for precipitates and solute segregation. Much corrosion is of an intercrystalline nature so that a coarse-grained structure tends to have a better corrosion resistance than a fine-grained one of the same metal.

c) Removal of the protective skin due to surface abrasions.

d) Differences due to thermal or mechanical effects.

e) Differences in composition of electrolytes, which often occur under stagnant conditions.

f) The presence of stress (including internal stress).

g) Differential aeration, which may have many sources, such as scale adhering to parts of the surface.

Arising from the above, corrosion cells (other than dissimilar electrode cells) may be broadly classified as:

1 Stress cells.
2 Differential temperature cells.
3 Concentration cells:
   a) Salt concentration cells.
   b) Differential aeration cells.

Stress Cells

The atoms of a highly stressed metal tend to ionise to a greater extent than atoms of the same metal in an unstressed condition. Therefore, the stressed metal will be anodic with respect to the unstressed metal. Consequently an electrochemical cell may be set up in a component in which the stress distribution is uneven. This also applies to an uneven distribution of internal stresses, e.g. residual stresses in a cold worked component.

Differential Temperature Cells

This type of cell can occur when pieces of the same material are at different temperatures in an electrolyte of the same initial composition. Such a situation may occur in heat exchanger units or with immersion heaters. Predicting the polarity of the electrodes is very difficult.

Concentration Cells

a) Salt concentration cells

When a metal is in contact with a concentrated electrolyte it will not ionise as much as when it is in contact with a dilute electrolyte. Therefore, if a piece of metal is in contact with an electrolyte of varying concentration, those portions in contact with dilute electrolyte will be anodic relative to the portions in contact with more concentrated electrolyte.

This type of electrochemical cell may be set up in situations involving flowing electrolytes in pipes.

b) Differential aeration cells

This type of cell is due to differences in oxygen concentration and is more common than salt concentration cells.

Fig. 9.3 Apparatus to illustrate differential aeration

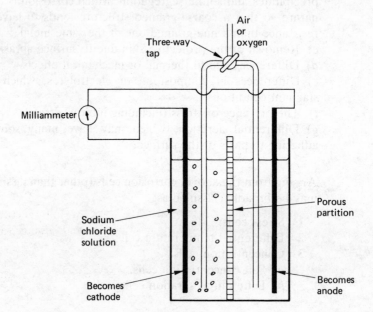

Three-way tap

Air or oxygen

Milliammeter

Porous partition

Sodium chloride solution

Becomes cathode

Becomes anode

The differential aeration effect may be demonstrated using the apparatus shown in fig. 9.3. Two identical pieces of the same metal are immersed in an electrolyte of dilute sodium chloride solution and the circuit completed as shown. No current is detected on the milli-ammeter because there is no difference in potential between the pieces of metal. When air or oxygen is passed into one compartment, a current is detected, the aerated metal (cathode) being protected while the other piece of metal (anode) corrodes. When the current of air or oxygen is diverted to the other compartment, the direction of the current is reversed. The differential aeration effect can be used to explain a number of corrosion problems, e.g. the rusting of iron under drops of water as represented in fig. 9.4. The differences in oxygen concentration between the centre and the periphery of the drop of water causes a corrosion cell to be set up. At the centre where the concentration of available oxygen is least the anodic reaction predominates, i.e.

$$Fe \rightarrow Fe^{2+} + 2e \tag{1}$$

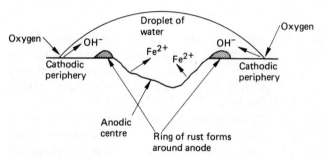

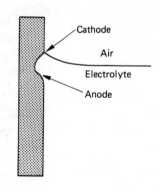

Fig. 9.4 Corrosion of iron under a water droplet

Fig. 9.5 Water-line attack caused by differential aeration

At the peripheral regions, the oxygen supply from the atmosphere maintains the cathodic reaction:

$$\tfrac{1}{2}O_2 + H_2O + 2e \rightarrow 2(OH)^- \tag{2}$$

Adding (1) and (2):

$$Fe + \tfrac{1}{2}O_2 + H_2O \rightarrow Fe(OH)_2$$

$Fe(OH)_2$ is rather unstable in aerated solution and is oxidised to form brown rust $Fe_2O_3.H_2O$.

The concentration of oxygen in a liquid usually decreases as the depth increases, so that the deepest parts of immersed equipment will be anodic to those nearer the water line.

A severe form of corrosion may occur at the water line of steel vessels which contain aqueous solutions (fig. 9.5). Differential aeration causes the corrosion which is worst just below the water line.

9.5 Types of Corrosion Attack

Two types of corrosion damage may be identified:

1 Uniform attack
2 Local attack.

Uniform Attack

This is the most common type of corrosion. A quantitative assessment may be expressed in terms of mass of metal lost, e.g. grammes per square metre per day. Uniform attack results in the heaviest masses of metal lost, but the rate of attack can usually be reliably determined so that the service life of components can be fairly accurately estimated. However, what is usually far more important than the amount of general surface corrosion is the pattern of the attack.

Local Attack

If the attack is localised the corrosion rate is exaggerated, being almost directly proportional to the ratio of the areas of cathode to anode. It follows that this type of damage is far more serious than uniform attack and may cause premature failure, such as the leaking of tanks. There are several types of local attack including the following:

Pitting is caused mainly by differential aeration or by the presence of stress. The initial depression in the surface may be due to a break in a protective film or the presence of scale. Once a pit is formed the corrosion proceeds rapidly because the surface of the metal (cathode) has greater access to oxygen (fig. 9.6) than the bottom of the pit (anode). Corrosion is accelerated because the surface area of the cathode is much greater than that of the anode. The corrosion product accumulates at the mouth of the pit and aggravates the corrosion by making access to oxygen more difficult.

Fig. 9.6 Pitting corrosion

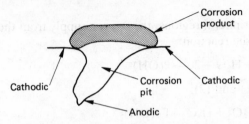

Intergranular corrosion occurs by localised attack at grain boundaries which behave as anodes to the larger surrounding cathodic areas. Incorrectly heat-treated austenitic stainless steels and duralumin-type alloys are prone to this type of attack, as are plain carbon steels in the presence of nitrate solutions.

Graphitisation is a form of selective attack on grey cast iron, resulting in a soft, powdery surface. Graphite is cathodic to ferrite, and the iron dissolves leaving a weak network of graphite.

Stress-corrosion cracking may be defined as the premature failure of a material due to the combined action of a static tensile stress and a corrosive environment. In stress-corrosion cracking there is usually very little overall corrosion, but the metal breaks by the passage of a macroscopically sharp and narrow crack across the axis of tensile stress.

Stress-corrosion cracking starts at a surface pit or notch such as a break in a protective film. The cracking may be intergranular or transgranular and the appearance of the fracture surfaces is similar to that after brittle fracture.

The earliest cases of stress-corrosion cracking was the so-called "season cracking" of brass cartridge cases in the nineteenth century. This was a type of brittle intergranular fracture which occurred when cold-worked α-brass, containing considerable internal stress, was exposed to atmospheres containing small amounts of ammonia. The trouble was overcome by annealing the brass, after working, at about 300°C which relieved the internal stress without unduly softening the brass.

Most metals are subject to stress-corrosion cracking when stressed in specific environments and the following examples have been identified:

a) Caustic embrittlement in boilers due to the combined action of high stress (e.g. at rivet holes) and large concentration of hydroxyl ions (from the soda-ash added to the boiler feed water).

b) Intergranular cracking of heat-treated aluminium alloys in the presence of chloride ions.

c) Intergranular cracking of mild steel in nitrate solutions.

d) Transgranular cracking of austenitic stainless steels in chloride solutions.

Corrosion Fatigue

When a metal is subjected to repeatedly applied stress in a corrosive environment, the fatigue resistance may be lowered, although the loss of metal by corrosion may be negligible. This is referred to as corrosion fatigue. [See also page 50 of *Structure*]. The phenomenon usually occurs by transgranular cracking and is most common in those environments which cause pitting. Under conditions conducive to corrosion fatigue, S-N curves for ferrous materials no longer show a fatigue limit but resemble the curves of non-ferrous materials. Water vapour in the atmosphere considerably reduces the fatigue strength of certain heat-treated aluminium alloys. In aggressive environments plain carbon steels appear to possess no fatigue limit, and given sufficient time, fail at very low stresses. For example, a 0.15% carbon steel tested in air showed a fatigue limit of $250 \, \text{N}/\text{mm}^2$ at 5×10^7 cycles but the value was lowered to $60 \, \text{N}/\text{mm}^2$ when tested in sea water. Fatigue life in corrosive media can be increased by care in design, the use of coatings, or introducing surface compressive stresses by shot peening which help to prevent the development of notches.

Hydrogen Damage

When atomic hydrogen is liberated at a cathode, some of the atoms unite to form molecules but others may diffuse into the metal in atomic form. At certain positions (e.g. voids) within the metal, molecular hydrogen can form and build up high internal pressures. This can result in blistering of the surface or embrittlement, and the latter may be a cause of stress-corrosion cracking.

9.6 Prevention of Aqueous Corrosion

In combating corrosion the importance of good design cannot be over-emphasised because this helps to reduce contact with the corrosive medium (fig. 9.7). Crevices which allow collection of moisture and dirt to form stagnant areas should be avoided, as should a design in which cleaning and maintenance are difficult. Badly ventilated spaces and poorly sealed joints obviously increase the opportunity of corrosive attack.

Methods of minimising or preventing aqueous corrosion may be classified as follows:

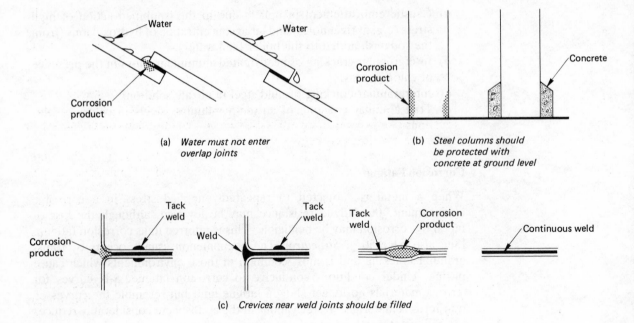

(a) *Water must not enter overlap joints*

(b) *Steel columns should be protected with concrete at ground level*

(c) *Crevices near weld joints should be filled*

Fig. 9.7 Effect of design on corrosion

1 Modification of the environment.
2 Modification of the metallic material under attack.
3 Cathodic protection.
4 Protective coatings.

Modification of the Environment

An aqueous environment may be modified by the removal of its oxygen or by addition of inhibitors. Examples of the above include de-aeration of boiler water and the use of inhibitors in car radiators.

Dissolved gases may be removed by increasing the temperature or by holding at low pressure and flushing with an inert gas such as nitrogen. Chemical de-aeration may be achieved by adding sodium sulphite Na_2SO_3 to alkaline solutions or hydrazine N_2H_4 to neutral or acid solutions.

Inhibitors are substances which, when added to a corrosive aqueous environment, stifle the corrosion reaction. Inhibitors may be classified in a simple way according to whether they act upon anodes or at cathodes. Inhibitors must be present in a minimum concentration to be effective.

Cathodic inhibitors form a protective film or precipitate at cathode regions and function in neutral or, in a few cases, alkaline solutions. Examples of this type of inhibitor include soluble salts of magnesium and zinc as well as calcium bicarbonate. An insoluble precipitate of oxide or hydroxide is formed on the cathode, e.g.

$$Mg^{2+} (aq) + 2OH^- (aq) \rightarrow Mg(OH)_2(s)$$

Such precipitates must be electrical insulators if they are to be effective, otherwise the cathodic reaction would continue on the outer surface of the film.

If the concentration of the inhibitor falls below the minium required it will be ineffective, but will not cause severe localised corrosion as is the case with anodic inhibitors.

Anodic inhibitors precipitate an insoluble compound at anodic regions and this stops the corrosion reaction. For example, in the case of steel, the Fe^{2+} ions as they leave the metal are converted into an insoluble compound which block the anodic reaction.

Typical substances used as anodic inhibitors include sodium chromate, sodium phosphate as well as various alkalis such as sodium hydroxide, sodium carbonate and sodium silicate.

Since anodic inhibitors have no direct effect on the cathodic reaction, there is a considerable risk of an intensive anodic reaction at localised weak spots, leading to pitting, if the inhibitor fails to seal off the anode completely. It is essential, therefore, that the concentration of anodic inhibitor should exceed the minimum amount needed to produce a complete film.

Some substances such as sodium benzoate appear not to produce pitting, whatever the concentration, and are used as anodic inhibitors in motor-car radiators.

Pickling restrainers   Hot-worked metal is often pickled in acid solution (e.g. H_2SO_4, HCl) in order to remove the mill scale before further processing of the metal. It is important that, once the scale has been removed, the attack of the acid on the bare metal is minimised and, for this purpose, restrainers are added to the pickling solution. These are organic substances such as pyridine, quinoline and thiourea which contain sulphur and nitrogen. The molecules of these substances attach themselves to the surface of the metal and so form an adsorbed layer which inhibits the attack on the metal. A further important requirement of pickling restrainers is that they should limit the uptake into the metal of hydrogen released by the acid attack, which can cause blistering of the metal and hydrogen embrittlement.

Modification of the Material

The corrosion resistance of many materials may be improved by *alloying*. Alloys are used which simulate noble metal behaviour by passive oxide film formation. Therefore, Al, Cr, Ti, Ni-Cr alloys and stainless steels are used for specific environments. Reducing conditions may cause oxide breakdown, but aerated solutions promote oxide self-healing. On the other hand, Ni-Cu alloys (e.g. Monel metal) which are only suitable for non-oxidising acids show best corrosion resistance in de-aerated acid solutions. Nickel and its alloys are very resistant to alkaline solutions. The resistance of mild steel to atmospheric attack may be increased by small additions of copper or chromium, which render the oxide film more protective. Corrosion resistance may also be obtained if an adherent insoluble film is formed on the surface of a metal, e.g. lead immersed in H_2SO_4 forms insoluble $PbSO_4$.

Fig. 9.8 Cathodic protection

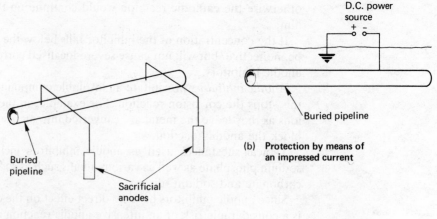

(b) **Protection by means of an impressed current**

(a) **Protection by means of sacrificial anodes**

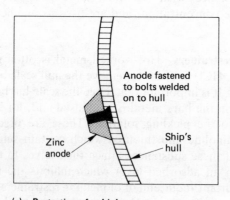

(c) **Protection of a ship's hull by sacrificial anodes**

Cathodic Protection

In this method of protection, galvanic cells are deliberately formed (fig. 9.8) and the material to be protected is made the cathode. Metals which are strongly anodic with respect to steel (e.g. Zn, Mg, Al) may be used as sacrificial anodes to prevent the corrosion of buried steel pipelines or for the protection of ships' hulls below the waterline.

The sacrificial anode maintains a supply of excess electrons to the material which is to be protected, thus ensuring that the latter is the cathode. The replacement of consumed anodes is usually an easy and relatively inexpensive operation.

Another way of arranging that the metal to be protected is cathodic is by applying a small d.c. voltage using an inert anode (e.g. graphite).

Protective Coatings

One of the commonest methods of minimising corrosion is to isolate the material from its environment by using a protective coating. Coatings may be metallic or non-metallic and afford protection either by excluding direct contact with the environment or by sacrificial cathodic protection. With all coatings proper surface preparation prior to the application of the coating is vital. Rust and scale may be removed by acid pickling or abrasive cleaning, while oil and grease can be eliminated by hot alkaline solutions or by organic solvents.

Metallic coatings A thin layer of a metal with good corrosion resistance may be used to protect another metal. The coating metal may be either anodic or cathodic with respect to the basis metal. With a cathodic coating, such as tin on steel, any discontinuity or flaw in the coating will set up a galvanic cell with the steel as the anode. In this case the corrosion of the steel will be accelerated by the galvanic action. On the other hand, in the case of an anodic coating, such as zinc on steel, any break in the coating will cause the zinc to corrode sacrificially and the steel will be protected until most of the zinc has corroded. The difference between the two types of protection is illustrated in fig. 9.9.

Fig. 9.9 Protection of steel by tin and by zinc

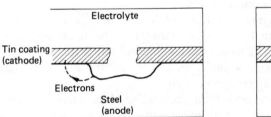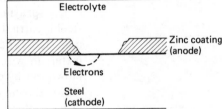

Metallic coatings may be applied by a variety of means including:

1 Hot dipping
2 Electrodeposition
3 Cladding
4 Spraying
5 Cementation

In **hot dipping**, the material to be protected is immersed in a molten bath of the coating metal (e.g. low melting point metals such as Zn, Sn, Pb or Al). Good adhesion to the basis metal may be obtained by the formation of an alloy layer at the coating-basis metal interface. However, coating-thickness control is poor and unnecessarily thick coatings are obtained. Hot dipping is now rarely used to produce coatings of costly metals, such as tin, and the method has been superseded by electrodeposition. However, the coating of large, complex-shaped steel components with zinc is still carried out by hot dipping (called galvanising).

In **electrodeposition** (electroplating), the component to be plated is made the cathode in an electrolytic cell. The electrolyte usually consists of a solution of a salt of the metal to be deposited, together with special additives.

The anode is usually made of the coating metal which dissolves into the electrolyte when the d.c. current is passed and plates out on the cathode. In a few cases, such as in chromium plating, an insoluble anode is used (e.g. Pb) and the metallic chromium is provided by the electrolyte.

This method is used to apply Cu, Zn, Cd, Ni, Cr, Ag and Au as thin, relatively pore-free coatings. The coating thickness is easily controlled and very thin coatings are possible. No alloying with the basis metal is involved and adhesion depends entirely on the intimate contact of the coating and basis metal.

In **cladding**, the metal to be protected is sandwiched between two pieces of the coating metal and then hot or cold rolled to give a cladding thickness of about 10% of the total thickness of the material. The coating metal and the basis metal must have similar deformation characteristics and some alloying takes place at the interface. A subsequent annealing (especially after cold rolling) may be necessary to cause greater diffusion at the interface and thus improve the bonding.

Duralumin coated with pure aluminium (i.e. "Alclad") is a well-known clad product, while "Niclad" (nickel-clad mild steel) is also commonly used. Clad steels are widely used in the chemical industry in order to avoid the use of thicker plates of expensive corrosion-resistant materials such as Monel metal and stainless steels.

The **spraying** of coatings of Al, Zn, Sn, Cu, Pb, brass, solder and stainless steel is carried out, the commonest use being the application of Zn and Al to mild steel. The molten metal is sprayed as droplets from a pistol, in which the metal, in wire form, is melted by electric arc or oxy-acetylene flame and blown out by compressed air.

No alloying occurs and the surface of the basis metal must be roughened in order to produce reasonable adhesion; subsequent annealing may be carried out in order to improve the adhesion.

The method is widely used because of the portability of the required equipment and flexibility of the technique; large structures such as bridges, buildings, storage tanks and pylons can be sprayed on site.

In **cementation** processes the protective metal is caused to diffuse into the surface of the component. The components are surrounded by the protective metal in powder form and heated to a sufficiently high temperature for diffusion to take place. The following are examples of cementation processes:

Sherardising in which a thin uniform coating of zinc is produced by heating the mild steel components in zinc dust at 350°C for several hours.
Calorising in which a layer of an iron-aluminium alloy is obtained by heating mild steel in powdered aluminium at about 1000°C.
Chromising in which a chromium-rich surface is obtained by heating mild steel in a powdered mixture containing chromium for a few hours at about 1400°C in an atmosphere of hydrogen, which is necessary to prevent oxidation of the chromium.

Non-metallic Coatings

The film of *oxide* which forms on a metal surface is sometimes very tenacious, dense and closely adherent, so that it will protect the underlying metal from corrosion and oxidation. For example, aluminium has good corrosion-resistance because of the hard, self-healing, adherent film of oxide on its surface.

In the **anodising** of aluminium alloys, the aim is to increase the thickness of this oxide film. This is achieved by making the component the anode in an electrolyte of chromic, sulphuric or oxalic acid, the cathode consisting of lead or stainless steel. The thickness, hardness and porosity of the anodised layer depends on the electrolyte used and the processing temperature. Porous anodised films can be coloured by immersion in a solution containing a suitable dye. Sealing of the pores of the anodised layer is essential in order to obtain good corrosion-resistance, and this can be achieved by immersion in boiling water or by the application of linseed oil.

In **phosphating**, the components are treated in, or with, an acid solution containing metal phosphates. Components to be treated are usually made of mild steel, but aluminium alloys and zinc-base alloys can also be phosphated.

Although the phosphate coating affords only very limited corrosion protection, it provides an excellent base for the retention of paint, varnish or oil. Phosphating is carried out on a variety of articles made from mild steel sheet (e.g. motor-car bodies, refrigerators, washing machines, office furniture) prior to painting.

Chromating produces oxide films on the surface of metals such as Al, Zn, Mg and their alloys. The components are immersed in a solution containing potassium dichromate and sulphuric acid. The colour of the resulting films depends on the bath composition and on the material being treated, and varies from yellow through grey to black. The treatment may be followed by painting.

Painting is the most familiar method of applying a protective coating. Oil paints consist of pigments suspended in linseed oil to which is added a thinner (white spirit or turpentine) and a drier. Pigments used include zinc oxide, zinc chromate, red lead, iron oxide, powdered aluminium and powdered zinc. Lead compounds are no longer widely used because of their poisonous nature. Some paints used on steel contain about 95% powdered zinc to further increase the corrosion resistance by sacrificial protection.

Paint films are electrical insulators so that the cathodic reaction cannot occur except at the metal surface. However, this reaction is not completely inhibited, because paint films are usually porous to oxygen and water which readily migrate through the film. To overcome this problem inhibitors (e.g. zinc chromate) are usually contained in the paint.

The effectiveness of a paint coating depends on the resistance it provides in the electrochemical path between anodic and cathodic areas of the metal as well as on the anodic inhibition it may provide. Fig. 9.10 illustrates the factors affecting the use of a paint system for protection.

Painting is a fairly cheap and convenient method of corrosion protection, but it has several disadvantages including the following:

Fig. 9.10 Factors affecting paint performance

Surface preparation	Duration of protection
Wire brushing	2 years
Pickling	10 years
Shot blasting	10 years

Effect of surface preparation on paint performance

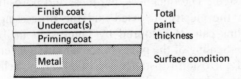

Finish coat	Total
Undercoat(s)	paint
Priming coat	thickness
Metal	Surface condition

a) A paint layer does not possess high wear resistance.

b) It deteriorates with time and periodic repainting is necessary.

c) Paint cannot be used for the protection of components which operate at elevated temperatures.

Organic coatings may be used including rubber, pitch, tar, bitumen, grease and thermoplastic materials. For example, the latter are widely used for covering steel wire trays in the manufacture of refrigerators and deep freezers.

Inorganic coatings used as environment-excluders include cements and vitreous enamels. The latter are rather brittle and easily damaged, but can be used at elevated temperatures (e.g. in the manufacture of ovens and cookers).

9.7 Atmospheric Corrosion

Most metallic materials when exposed to dry air at ordinary temperatures slowly become coated with a thin protective film and this is called *tarnishing*. When exposed to the atmosphere, however, such films are often liable to attack because of the impurities present in the air. Such impurities may include both solid and gaseous substances, e.g. carbon (soot), carbon compounds, water vapour, SO_2, H_2S, NH_3, NO_2, chlorides and metal oxides, depending on the geographical location.

In marine areas the air contains significant amounts of chlorides, whereas in industrial areas the atmosphere contains a relatively high proportion of SO_2, carbon and carbon compounds. The corrosion rate is also affected by climatic variables such as differences in temperature, rainfall, relative humidity and wind. The amount of moisture present is especially important and corrosion is more severe where the relative humidity is higher. However, high rainfall may wash away corrosive substances and hence may prove beneficial. Solid airborne impurities (e.g. carbon particles) are very important because they may absorb reactive gases to form an electrolyte with condensed water vapour. Of the gaseous contaminants, SO_2 is the most deleterious since it is converted into H_2SO_4 in the presence of moisture. When present in corrosion pits, this acid is very difficult to remove.

The *control of atmospheric corrosion* depends on the type of space which contains the atmosphere. Three types of space may be identified: outdoor, indoor, and small closed containers. Regarding outdoor conditions, the levels of gaseous pollutant (e.g. SO_2) can only be restricted within certain limits. Therefore the use of a protective coating or a more resistant material are the usual methods of combating corrosion. For indoors, air-conditioning may be used to provide a clean dust-free atmosphere with low relative humidity. In small enclosed spaces, a vapour phase inhibitor (VPI) may be used. A VPI is a soluble film-forming inhibitor with a low vapour pressure and usually consists of inhibitive anions of nitrite or benzoate together with organic cations. The rate of inhibitor evaporation must be adjusted so that it is neither quickly lost nor too slow to properly inhibit.

9.8 Corrosion in Soil

The important factors involved in corrosion occurring in soils are the moisture content, oxygen content, pH value, electrical conductivity, and bacterial content of the soil; these factors are in turn largely governed by the chemical composition of the soil and its particle size.

The moisture content of soils can vary between very wide limits, i.e. from a dry sandy type to waterlogged conditions. A dry soil will usually produce less corrosion than a very wet soil. A soil's ability to retain water and oxygen is governed mainly by the particle size, so that a coarse gravel usually retains more oxygen than a clay soil. In poorly aerated soils, atmospheric oxygen may be unable to come into contact with the metal surface so that a protective film does not form. In well aerated soils the initially formed corrosion product usually reduces the subsequent rate of corrosion.

Soils of different composition can produce differential aeration cells with the anodic and cathodic regions several kilometres apart—this effect is called longline corrosion. The conductivity and pH of a soil will be governed by the chemical composition of the soil. In general, the higher the conductivity, the more corrosive the soil. The pH of soils is usually about 5 to 8, but dissolved CO_2 may form H_2CO_3 and result in acid conditions.

Bacteria in the soil can also cause chemical changes, e.g. H_2SO_4 may be produced by the oxidation of sulphur-containing compounds and thus cause corrosion. Some anaerobic bacteria (i.e. bacteria capable of living in absence of free oxygen) can reduce sulphate ions to hydrogen sulphide:

$$SO_4^{2-} + 4H_2 \rightarrow S^{2-} + 4H_2O$$

Underground steel pipes can be severely corroded as a result of the above reaction—the FeS scale which is formed being completely unprotective. This type of attack takes place in neutral, water-logged clayey soils in which internal access of atmospheric oxygen is virtually impossible. The sulphur-containing compounds in the soil may include soluble compounds as well as insoluble ones (e.g. $CaSO_4$). This micro-biological corrosion of iron and steel is evident from the local blackening of the soil near the metal, as well as from the smell of H_2S which is sometimes prevalent.

Another cause of underground corrosion is "stray" electrical currents (a.c. or d.c.), for example where buried pipes are near electric railways.

To *control soil corrosion*, cathodic protection is often used, usually in combination with a coating of bitumen. Pipelines and storage tanks may be buried in sand of high electrical resistance.

Exercises 9

1   *a*) Explain what is meant by "dry corrosion".
    b) Name one metal which forms a protective film against dry corrosion.
    c) Name an element commonly added to steel to give protection against dry corrosion.

2   Write ionic equations to represent the reactions taking place at *a*) the anode and *b*) the cathode during the dry corrosion of iron by oxygen.

3   Explain what is meant by the following terms:
    a) the electrochemical series, *b*) an electrochemical cell.

4   Write ionic equations to represent two reactions which may take place at each of the electrodes in an electrochemical cell.

5   Explain what is meant by the terms:
    a) the galvanic series, *b*) passivity.

6   *a*) Name two types of electrochemical cell.
    b) Name two types of corrosion damage.

7   Explain what is meant by "stress-corrosion cracking" and give two examples of its occurrence.

8   Define the term "corrosion fatigue".

9   List four methods of minimising aqueous corrosion.

10 Make a labelled sketch of the apparatus you would use to give a demonstration of the "differential aeration effect" in corrosion.

11 *a*) Name four metals commonly used as coatings to protect steel from aqueous corrosion.
    b) Name four methods of applying a protective coating.

12 Write a brief account of the use of inhibitors to prevent aqueous corrosion.

13 Compare and contrast the use of electrodeposited tin and hot dipped zinc as methods of protecting steel from corrosion.

Index